Hydrogen in Semiconductors

MATERIALS RESEARCH SOCIETY
SYMPOSIUM PROCEEDINGS VOLUME 813

Hydrogen in Semiconductors

Symposium held April 13–14, 2004, San Francisco, California, U.S.A.

EDITORS:

Norbert H. Nickel
Hahn-Meitner-Institut Berlin
Berlin, Germany

Matthew D. McCluskey
Washington State University
Pullman, Washington, U.S.A.

Shengbai Zhang
National Renewable Energy Laboratory (NREL)
Golden, Colorado, U.S.A.

Materials Research Society
Warrendale, Pennsylvania

Single article reprints from this publication are available through
University Microfilms Inc., 300 North Zeeb Road, Ann Arbor, Michigan 48106

CODEN: MRSPDH

Published by:

Materials Research Society
506 Keystone Drive
Warrendale, PA 15086
Telephone (724) 779-3003
Fax (724) 779-8313
Web site: http://www.mrs.org/

Manufactured in the United States of America

CONTENTS

HYDROGEN IN ELEMENTAL SEMICONDUCTORS

*Invited Paper

*Invited Paper

GENERAL PROPERTIES OF HYDROGEN IN SEMICONDUCTORS

*Invited Paper

PREFACE

The study of hydrogen in solids is of significant importance in modern technology. Hydrogen, often present in large concentrations by the various growth techniques, affects the structural, electronic, and optical properties of the crystals. Isolated hydrogen is often electrically active, causing passivation or charge compensation of defects and impurities. Since the presidential proposal of the Freedom Car, hydrogen in semiconductors takes another twist as a potential source of clean fuel of the future. The study of hydrogen transport, diffusion, and chemical reaction in the solids is thus also of significant social importance.

The studies of hydrogen in the various semiconductors, despite their common concerns, often proceed in different directions. This is somewhat expected as hydrogen in small to medium gap materials like silicon or GaAs is amphoteric whereas recently, it was found that in larger gap materials such as ZnO, hydrogen could act only as a donor, as predicted by theory. As often happens in science, different groups of researchers studying in hydrogen in solids do not always speak the same language. For example, few people would consider silicon as a practical fuel storage material due to its weight, but in reality, aluminum hydrides have been seriously considered for such a purpose. The knowledge acquired in the last half century for hydrogen in conventional semiconductors has not been effectively transferred over to the study of hydrogen storage in emerging semiconductors such as carbon nanotubes. Symposium H, "Hydrogen in Semiconductors," held April 13–14 at the 2004 MRS Spring Meeting in San Francisco, California, brought together researchers from various research areas to encourage cross fertilization and wider dissemination of the advances in this important field of study. This volume contains the papers presented at the symposium.

Major highlights of the conference include recent developments in the understanding of light-induced metastability in amorphous silicon, hydrogen in zinc oxide as a means of n-type doping, and hydrogen doping of dilute nitrides that modifies the band gap in some rather surprising ways. Hydrogen interaction with donors and acceptors is a never-ending story that was discussed at this symposium. It has been speculated that hydrogen binds to nitrogen acceptors in ZnO, contributing to the failure to reliably produce p-type ZnO. The results for hydrogenation of boron-doped diamond are even more puzzling, as it appears to convert the conductivity from p-type to n-type. Even in well-established fields such as platelet formation, important new results have been reported. New theoretical insights have in some cases not just interpreted available experiments but also led the way for new research opportunities—a prime example being the exclusive donor behavior of hydrogen in ZnO.

We would like to thank the session chairs for their assistance in running the symposium and the referees who helped ensure the quality and clarity of the proceedings. We would especially like to thank every author who has contributed to the proceedings volume. We would also like to thank the MRS staff for their help in organizing and running the symposium to ensure its success. Above all, we would like to thank the symposium participants for their interesting and active work in the vibrant field of hydrogen in semiconductors.

Norbert H. Nickel
Matthew D. McCluskey
Shengbai Zhang

June 2004

MATERIALS RESEARCH SOCIETY SYMPOSIUM PROCEEDINGS

Volume 807— Scientific Basis for Nuclear Waste Management XXVII, V.M. Oversby, L.O. Werme, 2004, ISBN: 1-55899-752-0

Volume 808— Amorphous and Nanocrystalline Silicon Science and Technology—2004, R. Biswas, G. Ganguly, E. Schiff, R. Carius, M. Kondo, 2004, ISBN: 1-55899-758-X

Volume 809— High-Mobility Group-IV Materials and Devices, M. Caymax, E. Kasper, S. Zaima, K. Rim, P.F.P. Fichtner, 2004, ISBN: 1-55899-759-8

Volume 810— Silicon Front-End Junction Formation—Physics and Technology, P. Pichler, A. Claverie, R. Lindsay, M. Orlowski, W. Windl, 2004, ISBN: 1-55899-760-1

Volume 811— Integration of Advanced Micro- and Nanoelectronic Devices—Critical Issues and Solutions, J. Morais, D. Kumar, M. Houssa, R.K. Singh, D. Landheer, R. Ramesh, R. Wallace, S. Guha, H. Koinuma, 2004, ISBN: 1-55899-761-X

Volume 812— Materials, Technology and Reliability for Advanced Interconnects and Low-k Dielectrics—2004, R. Carter, C. Hau-Riege, G. Kloster, T-M. Lu, S. Schulz, 2004, ISBN: 1-55899-762-8

Volume 813— Hydrogen in Semiconductors, N.H. Nickel, M.D. McCluskey, S. Zhang, 2004, ISBN: 1-55899-763-6

Volume 814— Flexible Electronics 2004—Materials and Device Technology, B.R. Chalamala, B.E. Gnade, N. Fruehauf, J. Jang, 2004, ISBN: 1-55899-764-4

Volume 815— Silicon Carbide 2004—Materials, Processing and Devices, M. Dudley, P. Gouma, P.G. Neudeck, T. Kimoto, S.E. Saddow, 2004, ISBN: 1-55899-765-2

Volume 816— Advances in Chemical-Mechanical Polishing, D. Boning, J.W. Bartha, G. Shinn, I. Vos, A. Philipossian, 2004, ISBN: 1-55899-766-0

Volume 817— New Materials for Microphotonics, J.H. Shin, M. Brongersma, F. Priolo, C. Buchal, 2004, ISBN: 1-55899-767-9

Volume 818— Nanoparticles and Nanowire Building Blocks—Synthesis, Processing, Characterization and Theory, O. Glembocki, C. Hunt, C. Murray, G. Galli, 2004, ISBN: 1-55899-768-7

Volume 819— Interfacial Engineering for Optimized Properties III, C.A. Schuh, M. Kumar, V. Randle, C.B. Carter, 2004, ISBN: 1-55899-769-5

Volume 820— Nanoengineered Assemblies and Advanced Micro/Nanosystems, J.T. Borenstein, P. Grodzinski, L.P. Lee, J. Liu, Z. Wang, D. McIlroy, L. Merhari, J.B. Pendry, D.P. Taylor, 2004, ISBN: 1-55899-770-9

Volume 821— Nanoscale Materials and Modeling—Relations Among Processing, Microstructure and Mechanical Properties, P.M. Anderson, T. Foecke, A. Misra, R.E. Rudd, 2004, ISBN: 1-55899-771-7

Volume 822— Nanostructured Materials in Alternative Energy Devices, E.R. Leite, J-M. Tarascon, Y-M. Chiang, E.M. Kelder, 2004, ISBN: 1-55899-772-5

Volume 823— Biological and Bioinspired Materials and Devices, J. Aizenberg, C. Orme, W.J. Landis, R. Wang, 2004, ISBN: 1-55899-773-3

Volume 824— Scientific Basis for Nuclear Waste Management XXVIII, J.M. Hanchar, S. Stroes-Gascoyne, L. Browning, 2004, ISBN: 1-55899-774-1

Volume 825E—Semiconductor Spintronics, B. Beschoten, S. Datta, J. Kikkawa, J. Nitta, T. Schäpers, 2004, ISBN: 1-55899-753-9

Volume 826E—Proteins as Materials, V.P. Conticello, A. Chilkoti, E. Atkins, D.G. Lynn, 2004, ISBN: 1-55899-754-7

Volume 827E—Educating Tomorrow's Materials Scientists and Engineers, K.C. Chen, M.L. Falk, T.R. Finlayson, W.E. Jones Jr., L.J. Martinez-Miranda, 2004, ISBN: 1-55899-755-5

Prior Materials Research Society Symposium Proceedings available by contacting Materials Research Society

Hydrogen in Elemental Semiconductors

Hydrogen in Silicon and Germanium: Impurity Activation and Dopant Passivation

E. E. Haller

Department of Materials Science and Engineering, University of California, Berkeley and
Materials Sciences Division, Lawrence Berkeley National Laboratory, Berkeley, CA 94720 USA

ABSTRACT

The discovery of hydrogen-related deep and shallow level centers in ultra-pure germanium followed by hydrogen passivation of acceptors and donors in silicon marks the beginning of an exceptionally productive period of research. Both activation of neutral impurities and passivation of electrically active centers, including shallow and deep acceptors and donors, have been investigated with a wide variety of tools. The recent surge in interest in germanium by the silicon device community justifies a review of the effects of hydrogen in germanium, the first group IV semiconductor which exhibited signatures of electrically active hydrogen.

1. INTRODUCTION

Hydrogen is the most abundant element in our solar system, our galaxy, and the universe. Billions of suns fuse hydrogen into heavier elements, releasing the energy necessary to form and maintain life in locations where a number of stringent conditions are met. Our earth is one such location and we do not know with certainty where the next life-supporting planet resides! Hydrogen is a crucial ingredient of life in its combination with oxygen to form water and also in a very large number of organic and inorganic molecules. Hydrogen, a proton binding an electron, can accommodate a second electron in its 1s shell. Hydrogen can do this by forming a bond with another hydrogen atom, resulting in H_2, or with any other element or molecule which readily forms bonds. It should come as no surprise that hydrogen plays a wide variety of roles when combined with semiconductors. Most semiconductors form wide open lattices such as the diamond and the zincblende lattices or the wurzite lattice, real space structures which offer much space for the rather small hydrogen atom or the bare proton.

This review focuses mainly on the discovery and the characterization of hydrogen-containing impurity complexes which are electrically active. A number of early studies of hydrogen in semiconductors will be summarized. Emphasis will be given to germanium for several reasons. Activation and passivation of natural and electrically active impurities was discovered first in germanium. The development of ultra-pure germanium for gamma-ray spectroscopy offered unprecedented purity which enabled the discovery of hydrogen-containing centers at exceptionally low concentrations. Germanium currently experiences a revival in silicon device technology for two reasons. First, germanium has significantly higher mobilities of both electrons and holes than silicon, a factor which appears to become ever more important and second, alloying silicon with germanium can be utilized to introduce strain inside the device structure which, in turn, changes the bandstructure in beneficial ways. This revived interest in germanium justifies a "second" look at the effects of hydrogen in this semiconductor.

2. EARLY STUDIES OF HYDROGEN IN SEMICONDUCTORS

2.1. HYDROGEN PERMEATION STUDIES

Permeation of gas through a thin slab of a solid offers a rather direct way to measure the diffusivity and the solubility of this gas in the particular solid. The dependence of the solubility on the gas pressure difference across the solid allows us to draw conclusions about changes of the molecular state of the gas at the interface entering the solid. Van Wieringen and Warmoltz [1] were first in studying permeation of hydrogen and helium through silicon and germanium. They used crystalline thin wall semiconductor cylinders for their studies. They found the following key properties:

1) The solubility depends on the square root of the external hydrogen pressure $\left(\sqrt{p_{H_2}}\right)$ which means that the H_2 molecules break up into individual atoms upon entering the semiconductor.

2) The diffusivity is very large. For silicon, they found:

$$D_H(Si) = 9.4 \times 10^{-3} \exp\,(-0.48\ eV/k_BT)\ (cm^2 s^{-1}).$$

This result suggested that hydrogen occupies interstitial positions and that it is bound very weakly.

3) The solubility is rather low. At an external H_2 pressure of 1 atm, they found $\sim 10^{13}$ H (cm^{-3}) at 1200 °C. The temperature dependence of the solubility at 1 atm H_2 is given by:

$$S_H(Si) = 2.4 \times 10^{21} \exp\,(-1.86\ eV/k_BT)\ (cm^{-3}).$$

It is important to understand that these early studies were conducted with nominally undoped silicon and germanium and that at the temperatures used for the permeation experiments the material was intrinsic. We know from more recent studies that the solubility of hydrogen is a strong function of the dopant concentration present in a crystal. This has been observed for another interstitial impurity, the shallow lithium donor. Lithium forms neutral pairs with acceptors (lithium passivation) which, in turn, affects the total concentration of lithium dissolved in the crystal [2].

Using much improved experimental equipment, Frank and Thomas [3] performed hydrogen permeation studies with thin wall single crystal germanium cylinders. In close analogy to the van Wieringen and Warmoltz studies, they found a high diffusivity and a low solubility:

$$D_H\,(Ge) = 2.72 \times 10^{-3} \exp\,(-0.38\ eV/k_BT)\ (cm^2 s^{-1})$$

$$S_H\,(Ge) = 1.6 \times 10^{24} \exp\,(-2.29\ eV/k_BT)\ (cm^{-3}).$$

As in silicon, they found a $\sqrt{p_{H_2}}$ dependence of the solubility on pressure, indicating the break-up of H_2 molecules upon entry of hydrogen into germanium. In both studies, there is no mention of electrical activity of hydrogen.

2.2 HYDROGEN IN ZnO

In sharp contrast to the studies of hydrogen in silicon and germanium where no electrical activity was reported, Mollwo found that exposure of ZnO single crystals to H led to a reproducible increase of the n-type conductivity of these crystals [4]. This important finding lay dormant until its recent re-discovery. The current focus on ZnO as a wide bandgap semiconductor with a range of important potential applications has catapulted the interest in hydrogen in ZnO to the forefront. The latest findings on hydrogen in ZnO are reported in a number of papers published in these proceedings [5].

3. INTRODUCTION OF HYDROGEN INTO SEMICONDUCTORS: DELIBERATE *VERSUS* INCIDENTAL

3.1. CRYSTAL GROWTH AND PROCESSING

The chemical reducing properties of hydrogen are desirable in many bulk and thin film crystal growth processes. The ease with which hydrogen can be purified to extreme levels with a palladium permeation system is a further advantage of using hydrogen in crystal growth atmospheres. The choice of a 1 atm H_2 environment for ultra-pure germanium crystal growth [6] was based on these properties. What became a surprising observation over time was that all ultra-pure germanium crystals grown in a different atmosphere or in vacuum exhibited very poor charge collection when fashioned into p-i-n junction detectors. Today we know that passivation of deep level impurities and defects, which are mostly p-type in germanium, is the dominant reason why a H_2 atmosphere is a necessity when growing ultra-pure germanium single crystals used for gamma-ray detector applications.

The beneficial effects of exposure of finished semiconductor devices to forming gas (N_2 + a few % H_2), an age-old recipe in semiconductor processing, are most likely related to surface state and interface passivation. Hess et al. have shown that substitution of hydrogen with deuterium leads to a significantly slower hot electron degradation in silicon MOS field effect transistors [7]. In this case, hydrogen passivates SiO_2–Si interface states. The larger mass of deuterium reduces the probability for mechanical displacement by hot electrons.

3.2. PLASMA HYDROGENATION

The rapidly rising interest in the study of hydrogen in semiconductors after the discovery of electrical effects led to new, efficient introduction techniques. Exposure to H_2 led to the known low concentrations determined in the permeation experiments. One obvious solution to overcome the dissociation process of molecular to atomic hydrogen at the semiconductor surface was to dissociate molecular to atomic hydrogen. Formation of a hydrogen gas plasma was the obvious choice for a source of atomic hydrogen. Plasma discharges are always accompanied by copious photon production. If these photons can reach the semiconductor surface, they may change the local conditions significantly. Johnson [8] showed how the plasma generated energetic photons can be efficiently separated from the flux of hydrogen to the semiconductor surface (Fig. 1). Exposure to a hydrogen (or deuterium) plasma has become one of the most common hydrogenation techniques.

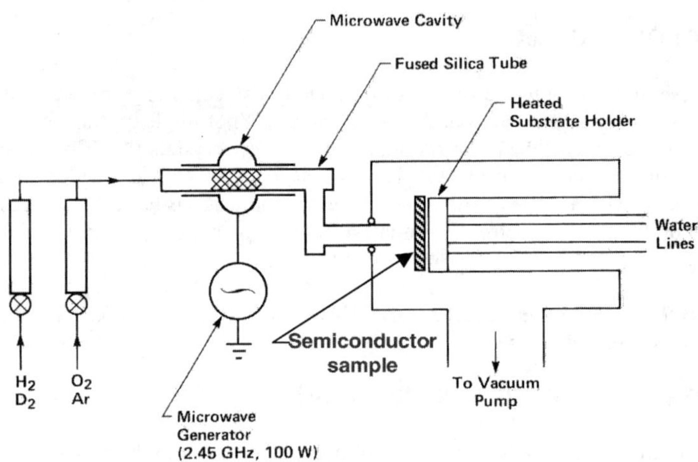

Figure 1. Plasma hydrogenation system (after Johnson [8]).

3.3. TRITIUM LABELING

Hydrogen does not only have a stable heavy isotope, deuterium, but also a radioactive isotope, tritium, with a nucleus consisting of one proton and two neutrons. The half-life of tritium is 12.26 years and it decays via β^- emission into $^3_2$He. Hansen et al. [9] grew an ultra-pure germanium single crystal in a hydrogen atmosphere containing tritium. The crystal was sectioned into small pieces whose exact location in the crystal was recorded. P-i-n junctions were formed and were operated with a large reverse bias at liquid nitrogen temperature in a "self-counting" mode until enough tritium decays had been recorded. No standards or calibrations were required for this experiment. The resulting hydrogen concentrations ranged from 5×10^{14} cm^{-3} to 2×10^{15} cm^{-3}, depending on the original location of the sample in the crystal.

A very similar experiment was conducted with the radioactive tracer element ^{14}C ($T_{1/2}$ = 5730 years). Up to that point in time, phase diagram compilations indicated that carbon was insoluable in germanium. It is not clear what this means at semiconductor impurity concentration levels. The ^{14}C experiments showed that carbon concentrations as high as 10^{14} cm^{-3} were present in ultra-pure germanium crystals grown from melts contained in a graphite susceptor [10]. Autoradiography pictures imaged the distribution of ^{14}C β^- decays on film and showed that some of the carbon resides in small carbon precipitates [11].

4. THE DISCOVERY OF ELECTRICALLY ACTIVE HYDROGEN IN GERMANIUM AND SILICON

4.1 THE DIVACANCY-HYDROGEN COMPLEX (V_2H) IN GERMANIUM

A surprising finding during the development of ultra-pure germanium was the observation that dislocation-free crystals showed severe charge trapping when made into p-i-n gamma-ray detectors. Variable temperature Hall effect measurements indicated the presence of an acceptor level at E_V + 80 meV [12]. Later on, Deep Level Transient Spectroscopy verified this result. Measurements of samples cut from partially dislocated crystals, one from the dislocation-free section next to one from the dislocated part of the crystal, showed the chemical impurities to be the same throughout one crystal slice. However, the sample from the dislocation-free area contained a deep acceptor at E_V + 80 meV (Fig. 2). A very interesting property of this acceptor is that its concentration can be changed reproducibly (within certain temperature limits!) from a few times 10^{10} cm^{-3} to ~ 3×10^{12} cm^{-3} (Fig. 3). The acceptor always showed up in hydrogen atmosphere grown dislocation-free crystals but never in the absence of hydrogen or in dislocated crystals. This observation strongly suggested the involvement of a native defect. The concentration of the new acceptor can be varied by heating to temperatures between 200 and 400 °C. If the new concentration is higher than the one existing in a sample, it is reached for all practical purposes instantaneously. Reducing the concentration by annealing at a lower temperature requires much longer times. This highly asymmetric behavior is characteristic for a dissociation-recombination reaction. Based on the combined experimental observations, we came up with the hypothesis that the acceptor in dislocation-free, hydrogen-grown germanium is a divacancy-hydrogen complex which can get fully passivated by binding a second hydrogen atom:

$$V_2H \ (E_V + 80 \ \text{meV}) + H \xrightarrow[\text{dissociation}]{\text{recombination}} V_2H_2 \ \text{(neutral)}$$

No spectroscopic information or isotope shifts have ever been obtained for the V_2H complex. This has made it impossible to construct a microscopic model for this defect.

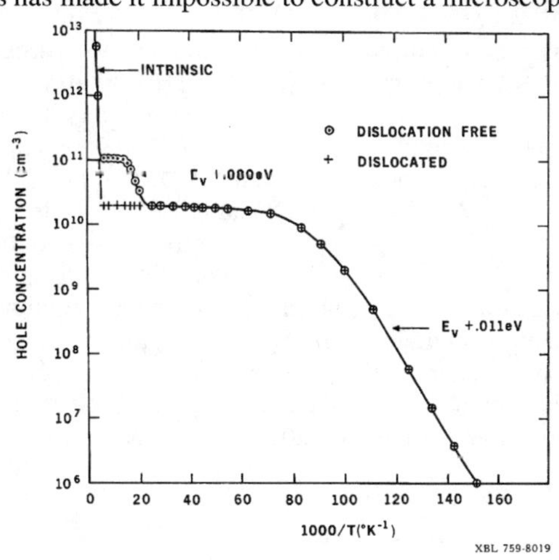

Figure 2. Log of the free hole concentration as a function of the inverse absolute temperature. The dislocation-free sample shows the E_V+80 meV acceptor identified as the V_2H center [12].

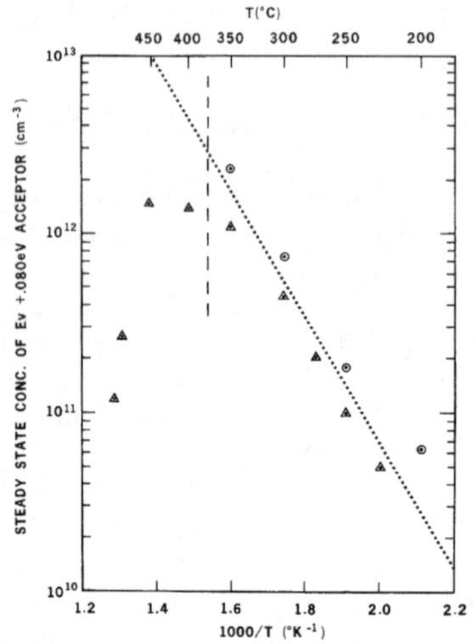

Figure 3. Steady state concentration of the V_2H acceptor as a function of the absolute inverse temperature for two dislocation-free samples from different crystals. The dashed vertical line indicates the temperature limit above which an irreversible loss of V_2H occurs. The dotted line is described by 1.04×10^{19} exp $(- 0.71$ eV/$k_BT)$ cm^{-3} [12].

Dislocation-free silicon does not appear to contain a similar defect. High resolution x-ray detectors made from ultra-pure, dislocation-free FZ silicon do not exhibit excessive charge trapping.

4.2. NEUTRAL IMPURITY ACTIVATION: A(H, Si), A (H, C), AND D (H, O) IN GERMANIUM

After rapidly quenching ultra-pure germanium crystals from ~ 425 °C, Hall discovered a shallow acceptor which annealed away near room temperature and a shallow donor which was slightly more stable [13,14]. An isotope shift in the ground states of the shallow acceptor and the shallow donor appearing in ultra-pure germanium crystals grown in a H_2 versus a D_2 atmosphere from a silica crucible was the very first unambiguous evidence that hydrogen was part of impurity complexes forming these shallow centers (Fig. 4) [15]. This 1978 discovery was followed up by a series of investigations which showed that silicon binding one hydrogen and forming a static, trigonal center led to this shallow acceptor [A(H, Si)] [16] and oxygen binding one hydrogen, which in turn was tunneling rapidly between four equivalent interstitial positions, formed the shallow donor [D(H, O)] [17]. A second acceptor was formed in graphite crucible-grown crystals and was modeled with a static, trigonal C-H complex [A(H, C)] [16]. That only one hydrogen participates in the formation of A(H, Si), A(H, C) and D(H, O) was determined by growing in pure and mixed H_2 and D_2 atmosphere (Fig. 5) [18]. The ground states of the two hydrogen containing acceptors are split. Splitting of a fourfold degenerate acceptor ground state requires an extra degree of freedom. This finding led to the early, incorrect assumption that the complexes had to be dynamic with tunneling hydrogen [19]. By contrast, the static, trigonal center model is simpler and it also explains the properties of partially passivated multivalent acceptors. Theoretical work by Denteneer et al. supports the experimental observation of the activation of neutral impurities by hydrogen in germanium [20]. The activation of neutral impurities has, to the knowledge of this author, remained unique. This hydrogen-related effect has not been found in silicon or the many III-V and more recently studied II-VI semiconductors.

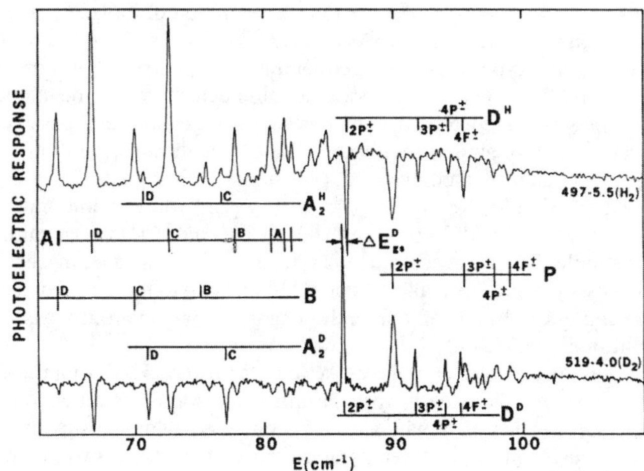

Figure 4. Spectra of a hydrogen grown p-type (# 497-5.5) and a deuterium grown n-type (519-4.0) ultra-pure germanium crystal showing the acceptor A(H,Si) and the donor D(H,O), and the acceptor A(D,Si) and the donor D(D,O), respectively. Both hydrogen and deuterium-related shallow centers display isotope shifts in their ground state binding energies [15].

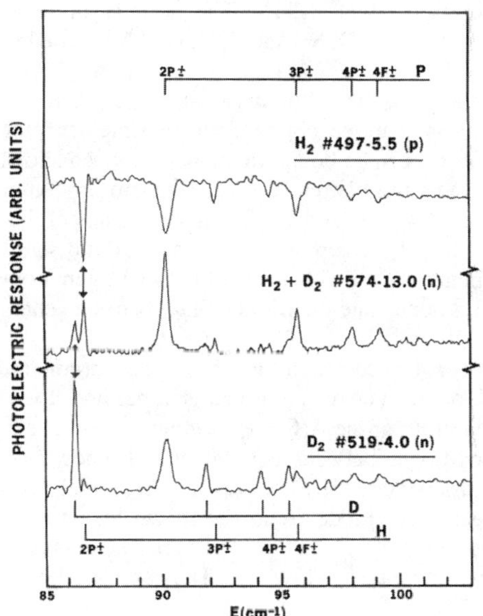

Figure 5. Growth in pure and mixed H_2/D_2 atmospheres indicates that only one H atom is involved in the formation of the donor D(H,O) [18].

4.3 PARTIAL PASSIVATION OF MULTIVALENT ACCEPTORS

Passivation of an electrically active acceptor or donor involves a short range pairing reaction. It is a process quite different from the commonly known effect of compensation in which donors and acceptors remain in their own locations. Passivation of dopant impurities leads to an increase in the free carrier mobility because the concentration of ionized scattering centers decreases. Compensation leads to a mobility decrease.

Passivation of acceptors in p-type silicon and germanium with interstitial lithium donors enjoyed great prominence during the time GeLi and SiLi radiation detectors were developed in the 1960s [21]. In most p-type semiconductors, the amphoteric hydrogen acts as a passivating donor and we should expect a behavior similar to lithium. The classic discovery of boron passivation by Sah et al. [22] and later by Pankove et al. [23] supports this notion.

What kind of interactions should one expect to occur between hydrogen and multivalent impurities such as double acceptors Be and Zn or even the triple acceptor Cu in germanium? The answer is surprisingly simple: Double acceptors bind one or two hydrogen atoms, each hydrogen reducing the valency by one. The triple acceptor Cu binds up to three hydrogen atoms. In germanium crystals containing Cu, Li and H, one finds copper acceptors partially passivated by one hydrogen and one lithium atom [24].

We discovered the shallow acceptor complexes A(Be,H) and A(Zn,H) during the development of sensitive far infrared photoconductors for infrared astronomy applications [25]. Crystals grown in a H_2 atmosphere and doped with Be or Zn contained unknown shallow acceptors with split ground states. Annealing at temperatures above 600 °C for several hours completely removed these novel acceptors and only the known chemical shallow acceptors plus the Be or Zn dopants remained. Detailed far infrared spectroscopy studies using uniaxial stress revealed trigonal Be-H and Zn-H complexes oriented along <111>. The partially passivated double acceptors exhibit properties similar to those of A(H, Si) and A(H, C). A detailed study has been conducted and reported by Kahn et al. [16].

The complexity of the structure of partially passivated multivalent centers increases when going from double to triple acceptors. As mentioned above, copper forms a triple acceptor in its substitutional form [26]. That two hydrogen atoms can be bound to copper has been shown by growing germanium crystals in H_2, D_2 and mixed atmospheres. Exposure to tritium-containing plasma allowed us to form monovalent complexes with H and T. The ground states of these various centers are markedly different. The A(Cu, H_2) acceptor shows a rich ground state manifold with a dozen closely spaced components. When either one of the hydrogen atoms is replaced by a heavier isotope D or T, this rich ground state manifold collapses to one energy state (Fig. 6).

Uniaxial stress spectroscopy studies reveal that the A(Cu, H_2) center has tetrahedral symmetry while all the centers containing one or more heavy hydrogen isotopes are of lower symmetry. In order to explain the 1s states manifold of the A(Cu, H_2) centers, Kahn et al. proposed a model involving tunneling of the hydrogen between equivalent real space positions. Rapid tunneling recovers the observed tetrahedral symmetry. The Devonshire model [27] can be used to explain qualitatively the observed effects. The model treats the energy levels of a hindered rigid rotor moving in a potential of octahedral symmetry. The motion changes from rotational to librational as the moment of inertia increases. The A(Cu, H_2) complex can be described as a rotor in a tetrahedral potential.

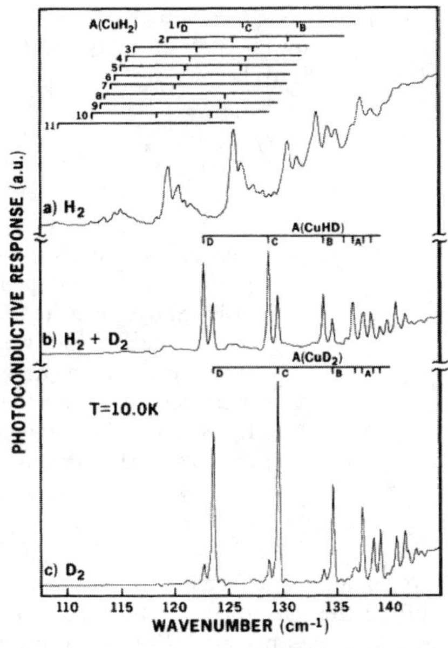

Figure 6. Photothermal ionization spectra of copper-dihydrogen acceptors in samples grown in different ambients. (a) Pure H_2, containing A(CuH$_2$), (b) 1:1 mixture of H_2 and D_2, containing A(Cu,H$_2$), A(CuHD), and A(CuD$_2$) in a 1:2:1 ratio, (c) Nearly pure D_2, containing A(CuD$_2$) and a trace of A(CuHD) (after Kahn et al. [38]).

The zero-point motion of hydrogen is assumed to be faster than the infrared transitions, leading to the observed recovery of tetrahedral symmetry. Substitution with a heavier hydrogen isotope sufficiently raises the moment of inertia, reducing rotational motion to librational motion and the tetrahedral symmetry can no longer be recovered.

Before closing this section, it is worth noting that a partially hydrogen passivated tunneling center has also been found in silicon. Muro and Sievers [28] showed that H or D transform the double acceptor beryllium into a single acceptor A(Be, H) or A(Be, D), respectively. The tunneling hydrogen or deuterium model developed first for germanium [19] quantitatively describes all major ground state and bound excited state properties of these partially passivated centers.

4.4. FULL HYDROGEN PASSIVATION OF ACCEPTORS AND DONORS IN SILICON AND III-V SEMICONDUCTORS

It cannot be the purpose of this brief survey to give a comprehensive account of all the research performed in this field. The focus here is on the early developments which led to several important discoveries and which form a foundation for the work which followed. Indeed, it took a surprisingly long time from the discovery of the divacancy-hydrogen center in germanium in 1976 [12] and the discovery of the isotope shift in the ground state of A(H, Si) and D(H, O) in 1978 [15] to the observation of boron passivation by hydrogen introduced from the surface of a boron doped silicon wafer by plasma exposure near 100 °C. [23]. The experimental evidence was the increase in resistivity measured by the spreading resistance technique (Fig. 7). Observation of local vibrational modes of the neutral B-H and B-D centers [29] and all the other group III acceptor-H complexes established the structure of the passivated acceptor complexes

[30]. The widely accepted structure of the acceptor-hydrogen complex consists of a hydrogen residing near a bond center position between the shallow acceptor impurity and a neighboring silicon atom, bound mainly to the silicon atom (Fig. 8) [31].

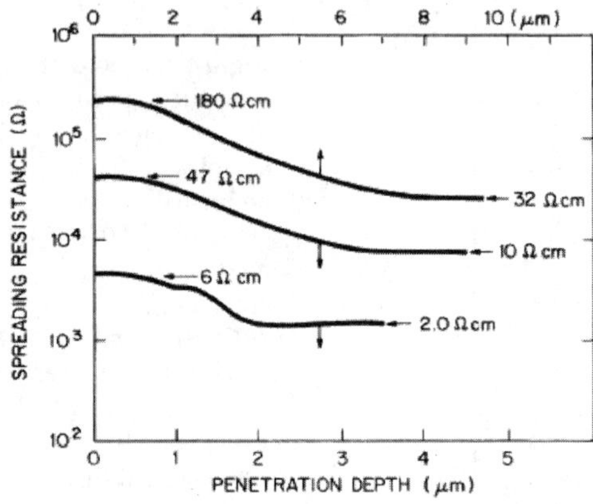

Figure 7. Spreading resistance profile of three B-doped samples of (100) Si hydrogenated for 1 h at 122 °C. The resistivities were obtained from a calibration curve. Note the great penetration depth of atomic hydrogen as the boron concentration decreases (after Pankove et al. [23]).

Shallow donor passivation by hydrogen was discovered by Johnson et al. [32]. Hall effect measurements showed a decrease in the free electron concentration and an increase in electron mobility after exposure of n-Si to a hydrogen plasma at 130 °C. These were the clear signs of hydrogen passivation, not of compensation of donors by an acceptor. Subsequent LVM spectroscopy studies led to a model of passivated donors which looks similar to the acceptor model with the important difference in the location of the hydrogen atom. It is located in an antibonding position bound to a silicon atom neighboring the donor atom (Fig. 9).

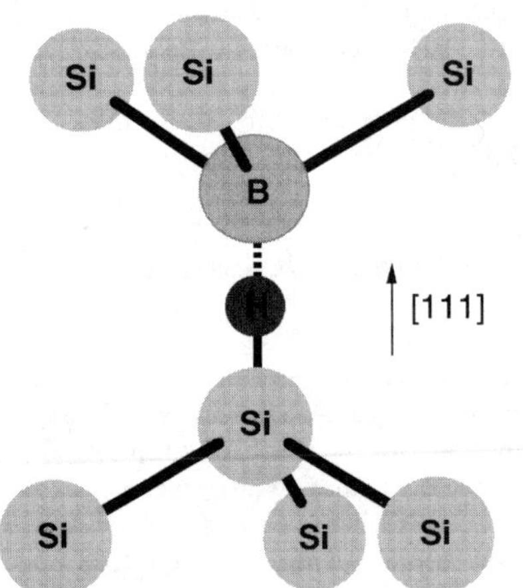

Figure 8. Model of a hydrogen passivated shallow boron acceptor in silicon.

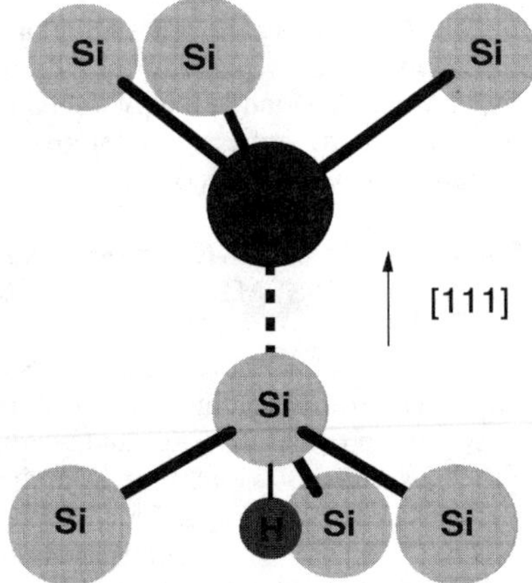

Figure 9. Model of a hydrogen passivated shallow arsenic donor in silicon.

In full analogy to germanium and silicon, hydrogen passivation of acceptors, donors and deep level defects was discovered and thoroughly studied in numerous III-V semiconductors [33]. A typical model of a hydrogen passivated donor in GaAs is shown in Fig. 10 [34].

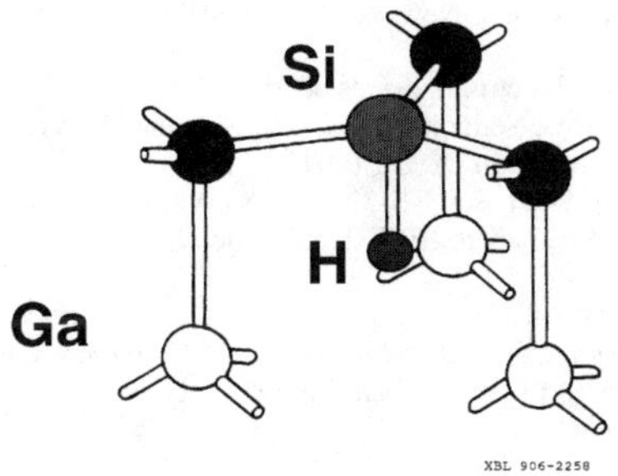

Figure 10. Model of the hydrogen passivated Si_{Ga} donor in GaAs (after Pajot et al. [34])

These were the beginnings of a vast amount of studies of hydrogen in semiconductors. The interested reader is referred to comprehensive surveys in books and proceedings [5, 35-37].

5. CONCLUSIONS

Looking back from today's perspective, it is clear that the development of ultra-pure germanium played an important role in the discovery of many hydrogen-related effects, including activation and passivation of electrically neutral and active impurities. It is quite ironic that the primary reasons for choosing a hydrogen atmosphere for crystal growth (chemically reducing and high purity) became secondary reasons after the discovery of deep level passivation by hydrogen. Ultra-pure germanium must contain hydrogen to be suitable for the fabrication of wide depletion layer p-i-n junctions working as gamma-ray spectrometers. Thanks in large part to hydrogen passivation of carrier traps, free holes and electrons have effective mean free paths of over 100 cm in high quality, ultra-pure germanium crystals at liquid nitrogen temperatures.

The studies of hydrogen in semiconductors have been remarkably productive for the past 30 years and many interesting results remain to be found in the future. Our detailed knowledge of hydrogen in semiconductors may contribute in unexpected ways to the development of the hydrogen economy!

ACKNOWLEDGMENTS

It is my privilege and pleasure to thank many of my former students and colleagues for the numerous productive collaborations we enjoyed, studying the effects of hydrogen in semiconductors. This work was supported in part by U.S. NSF Grant No. DMR-0109844 and by the Director, Office of Science, Office of Basic Energy Sciences, Division of Materials Science and Engineering, of the U.S. Department of Energy under contract No. DE-AC03-76F00098.

REFERENCES

1. A. van Wieringen and N. Warmoltz, Physica **XXII**, 849 (1956).
2. H. Reiss, C.S. Fuller, and F.J. Morin, Bell Syst. Tech. J. **35**, 535 (1956).
3. R.C. Frank and J.E. Thomas, Jr., J. Phys. Chem. Solids **16**, 144 (1960).
4. E. Mollwo, Z. Physik **138**, 478 (1954).
5. These proceedings: *Hydrogen in Semiconductors,* edited by N.H. Nickel and M.D. McCluskey, (Mater. Res. Soc. Proc. **813**, Warrendale, PA).
6. W.L. Hansen and E.E. Haller in *Nuclear Radiation Detector-Materials*, edited by E.E. Haller, H.W. Kraner and W.A. Higinbotham,(Elsevier Science Publishing Co., Inc., New York), Mater. Res. Soc. Proc. **16**, 1 (1983).
7. K. Hess, K.C. Kizilyalli and J.W. Lyding, IEEE Trans. Electron Devices **45**, 406 (1998).
8. N.M. Johnson, Chapter 7 in *Hydrogen in Semiconductors,* Semiconductors and Semimetals, Vol. **34**, edited by J.I. Pankove and N.M. Johnson, (Academic Press, Inc., San Diego, 1991), p. 113.
9. W.L. Hansen, E.E. Haller and P.N. Luke, IEEE Trans. Nucl. Sci. **NS-29**(1), 738 (1982).
10. E.E. Haller, W.L. Hansen, P. Luke, R. McMurray and B. Jarrett, IEEE Trans. Nucl. Sci. **NS-29**(1), 745 (1982).
11. P.N. Luke and E.E. Haller, J. Appl. Phys. **59**, 3734 (1986).
12. E.E. Haller, G.S. Hubbard, W.L. Hansen and A. Seeger, *Proceedings International Conerence on Radiation Effects in Semiconductors, 1976*, edited by N. B. Urli and J. W. Corbett, Inst. Phys. Conf. Ser. **31**, 309 (1977).
13. R.N. Hall, IEEE Trans. Nucl. Sci. **NS-21**(1), 260 (1974).
14. R.N. Hall, Inst. Phys. Confr. Series **23**, 190 (1975).
15. E.E. Haller, Phys. Rev. Lett. **40**, 584 (1978).
16. J.M. Kahn, R.E. McMurray, Jr., E.E. Haller and L.M. Falicov, Phys. Rev. B **36**, 8001 (1987).
17. B. Joos, E.E. Haller and L.M. Falicov, Phys. Rev. B **22**, 832 (1980).
18. E.E. Haller, W.L. Hansen and F.S. Goulding, Adv. Phys. **30**, 93 (1981).
19. E.E. Haller, B. Joos and L.M. Falicov, Phys. Rev. B **21**, 4729 (1980).
20. P.J.H. Denteneer, C.G. Van de Walle and S.T. Pantelides, Phys. Rev. Lett. **62**, 1884 (1989).
21. A.J. Tavendale and G.T. Evan, Nucl. Instrum. Methods **25**, 185 (1963).
22. C.T. Sah, J.Y.C. Sun and J.J. Tzou, Appl. Phys. Lett. **43**, 204 (1983); J. Appl. Phys. **55**, 1525 (1984).
23. J.I. Pankove, D.E. Carlson, J.E. Berkeyheiser and R.O. Wance, Phys. Rev. Lett. **51**, 2224 (1983).
24. E.E. Haller, G.S. Hubbard and W.L. Hansen, IEEE Trans. Nucl. Sci. NS-**24**(1), 48 (1977).
25. N.M. Haegel, E.E. Haller and P.N. Luke, Int. J. Infrared and Millimeter Waves **4**, 945 (1983).
26. R.N. Hall and J.H. Racette, J. Appl. Phys. **35**, 379 (1964).
27. A.F. Devonshire, Proc. Royal. Soc. London, Ser. A **153**, 601 (1936).
28. R. Muro and A. J. Sievers, Phys. Rev. Lett. **57**, 897 (1986).
29. N.M. Johnson, Phys. Rev. B **31**, 5525 (1985).
30. M. Stavola, S.J. Pearton, J. Lopata and W.C. Dautremont-Smith, Appl. Phys. Lett. **50**, 1086 (1987); Phys. Rev. B **37**, 8313 (1988).
31. K. Bergman, M. Stavola, S.J. Pearton and T. Hayes, Phys. Rev. B **38**, 9643 (1988).
32. N.M. Johnson, C. Herring and D.J. Chadi, Phys. Rev. Lett. **58**, 2077 (1986).
33. J. Chevallier, B. Clerjaud and B. Pajot, Chapter 13 in *Hydrogen in Semiconductors*, Semiconductors and Semimetals, Vol. **34**, edited by J.I. Pankove and N.M. Johnson, (Academic Press, Inc., San Diego, 1991), p. 447.
34. B. Pajot, *Shallow Impurities in Semiconductors 1988*, edited by B. Monemar, Inst. Phys. Conf. Ser. **95**, 437 (1989).
35. *Hydrogen in Semiconductors,* Semiconductors and Semimetals, Vol. **34**, edited by J.I. Pankove and N.M. Johnson, (Academic Press, Inc., San Diego, 1991).
36. *Hydrogen in Semiconductors II*, Semiconductors and Semimetals, Vol. **61**, edited by N.H. Nickel, (Academic Press, Inc., San Diego, 1999).
37. *Hydrogen in Materials and Vacuum Systems*, AIP Conf. Proc. **671**, edited by G.R Myneni and S. Chattopadhyay, (AIP, Melville, NY, 2003).
38. J.M. Kahn, L.M. Falicov and E.E. Haller, Phys. Rev. Lett. **57**, 2077 (1986).

Deciphering the Vibrational Spectrum of Interstitial H_2 in Si

Michael Stavola, E Elinor Chen, and W. Beall Fowler
Department of Physics and Sherman Fairchild Laboratory, Lehigh University, Bethlehem,
Pennsylvania 18015, USA

ABSTRACT

H_2 is a fascinating molecule whose properties revealed the influence of nuclear spin on the
molecular wave function in the 1920s. As an interstitial defect in Si, the H_2 molecule has given
rise to a number of perplexing puzzles since the discovery of its vibrational spectrum. The
absence of an ortho-para splitting for the H_2 vibrational line and an apparent low symmetry
found in stress experiments misled several researchers into thinking that interstitial H_2 in Si must
have a barrier to rotation. Our discovery of a new vibrational line for HD in Si and its
interpretation, along with the recognition that certain transitions are possible for HD, but not for
H_2 or D_2, establish that H_2 in Si is a nearly free rotator after all. Additional puzzles such as the
anomalous intensity of the HD line, the absence of an isotope dependence for the uniaxial stress
splitting of the H_2 and D_2 vibrational lines, and the properties of an O-H_2 complex are also
explained naturally. Recent Raman studies confirm that interstitial H_2 in Si is a free rotator but
raise interesting new questions about the diffusivities of the ortho and para species.

INTRODUCTION

The importance of interstitial H_2 molecules in semiconductors was suggested by theoretical work
performed in the early 1980s when interest in the properties of H in semiconductors was
beginning to grow [1-3]. The formation of H_2 molecules was also suggested to explain the
diffusion of H into Si and Ge [4]. Nonetheless, the H_2 molecule in a semiconductor was not
observed directly until recently when vibrational lines for the H_2, HD, and D_2 molecules in GaAs
[5] and Si [6-8] were discovered. In addition to the isolated H_2 defect, an O-H_2 complex is
formed when the H_2 molecule becomes trapped near an oxygen impurity in Czochralski-grown
Si [7]. Interstitial H_2 and O-H_2 in Si are shown schematically in Fig. 1. These experimental
results motivated a number of theoretical studies of the microscopic properties of interstitial H_2
in semiconductors [9-14].

For several years, it was not possible to reconcile a growing body of experimental results for
interstitial H_2 in Si with the properties observed for H_2 in GaAs or with the predictions of theory.
The Raman band of the H_2 molecule in GaAs is split into two components, 8 cm^{-1} apart, with an
intensity ratio of ~ 3:1 (ref. 5). These lines were assigned to ortho- and para-H_2, whose
frequencies differ because of ro-vibrational coupling. This interpretation leads naturally to the
conclusion that H_2, sitting at a T_d interstitial site in GaAs, is freely rotating. For the Si host,
several theoretical calculations for H_2 at a tetrahedral interstitial site found that <100>, <111>,
and <110> orientations have similar energies, suggesting that interstitial H_2 in Si should also
rotate nearly freely [9-12]. Furthermore, molecular dynamics calculations indicate that the H_2,
HD and D_2 molecules in Si behave as nearly free rotators, bouncing within the interstitial region
[13]. This conclusion is supported by our own model calculations for H_2 in Si [14] which
suggest a rotational barrier of only ~ 0.01 eV.

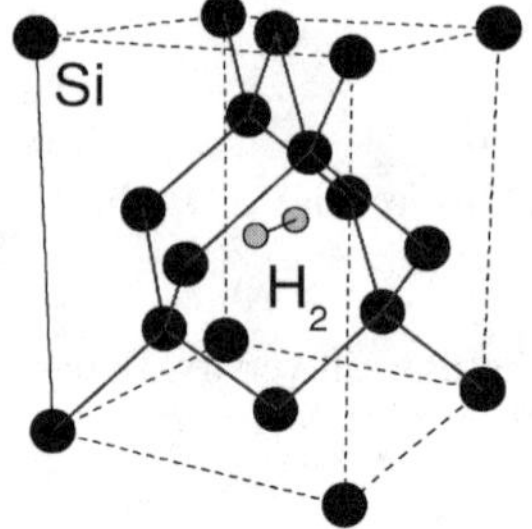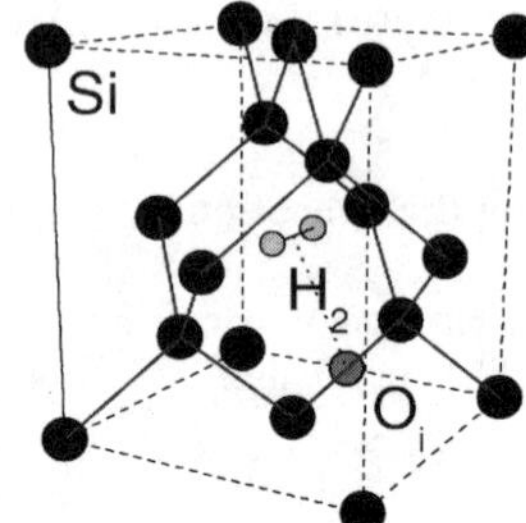

Figure 1. Interstitial H_2 (left) and the O-H_2 complex (right) in Si.

In contrast to the experimental situation in GaAs and the (naive) expectations inferred for H_2 in Si from theoretical results, the H_2 molecule in Si was found to give only a single H_2-vibrational line at 3618.4 cm^{-1} and no evidence for an ortho-para splitting in its Raman or infrared (IR) absorption spectra [6,8,12]. To explain the absence of an ortho-para splitting, it was suggested that there must be a barrier that prevents rotation of the molecule [12]. Uniaxial stress results for the 3618.4 cm^{-1} line of interstitial H_2 in Si were interpreted in terms of an orientationally degenerate defect with low symmetry, reinforcing the suggestion that the H_2 molecule is static [15]. While experiment seemed to suggest a static H_2 defect, this conclusion was in direct conflict with the predictions of theory and the rotational motion observed for H_2 in GaAs.

Additional problems with the interpretation of spectroscopic results for HD in Si were forthcoming. In addition to the 3618.4 cm^{-1} line for H_2 in Si, vibrational lines due to HD and D_2 were discovered at 3265.0 and 2642.6 cm^{-1} (4.2K), respectively, by IR absorption [6, 7] and Raman [8] spectroscopies. Spectra for the vibrational lines of H_2, HD, and D_2 in Si are shown in Fig. 2. In Si that contains both H and D, the HD vibrational line was reported to be weaker (by a factor of 7.5), relative to the intensities of the H_2 and D_2 lines, than would be estimated for a random pairing of H and D in the sample [16]. And the position of the 3265.0 cm^{-1} HD line lies 85 cm^{-1} to higher frequency than is predicted from the observed positions of the H_2 and D_2 lines and a simple model of the anharmonicity of the molecular vibration [15]. There are similar problems with the intensities and positions of the HD lines of the O-HD complex.

Our subsequent experimental results for the HD molecule and their interpretation show that interstitial H_2 in Si is actually a nearly free rotator after all [17, 18]. This conclusion leads to the solution of most of the perplexing puzzles associated with H_2 and the O-H_2 complex [19] in Si. Furthermore, the stretching modes of both ortho- and para-H_2 have been observed in recent Raman experiments [20]. However, these Raman experiments for H_2 in Si have uncovered new questions about the diffusivity of ortho- and para-H_2 in Si. The present paper is a survey of the solution to the problems surrounding interstitial H_2 in Si that has resulted from the deciphering of the H_2 vibrational spectrum.

KEY TO UNDERSTANDING H_2 IN Si: THE VIBRATIONAL MODES OF HD

The properties of the H_2 molecule are determined by the symmetry requirements of its total wave function [21]. H_2 consists of two identical nuclei with nuclear spin ½. If the nuclear spins are antiparallel, the nuclear wave function has odd parity, and, therefore, the rotational wave

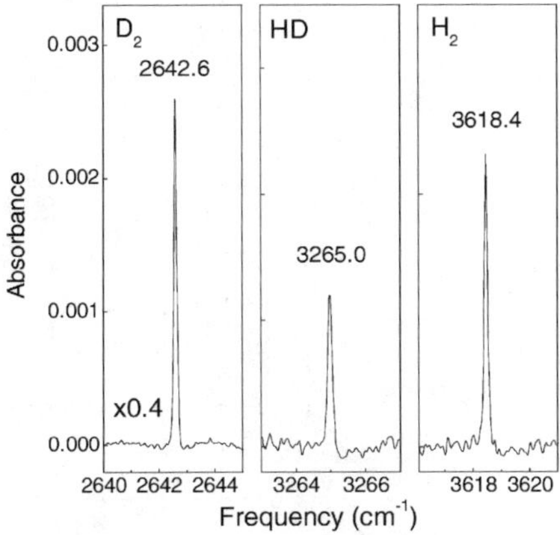

Figure 2. IR absorption lines (4.2K) assigned to interstitial D_2, HD, and H_2 in Si.

function must have even parity (with even values of the rotational quantum number J). If the nuclear spins are parallel, the nuclear wave function is even, and, therefore, the rotational wave function must be odd (with odd values of J). The nuclear spins are usually coupled only weakly to their environment, so the nuclear spin state can remain unchanged for very long times. These considerations result in there being two separate H_2 species, ortho and para, with even and odd nuclear-spin states, respectively, that do not thermalize on the time scale of many experiments. The ortho and para H_2 species have different rotational levels and spectra. D_2 behaves similarly except that in this case, because the nuclear spin of a deuteron is 1, the total molecular wave function must be symmetric. HD, however, is different because the nuclei are no longer identical, and the symmetry of the nuclear spin state does not restrict the rotational states that are possible.

The considerations above, valid for a freely rotating H_2 molecule, had been thought to be irrelevant for H_2 in Si because distinct ortho and para species were not observed in the vibrational spectra of the defect. On the contrary, the following results for HD in Si show the importance of molecular rotation. Fig. 3 shows spectra, measured as a function of temperature, for Si that contained H and D [17]. As the temperature is increased, the 3265.0 cm⁻¹ line previously assigned [6] to isolated HD becomes weaker while a new line at 3191.1 cm⁻¹ grows in with an activation energy of 71 ± 4 cm⁻¹. The difference in frequency between the 3265.0 and 3191.1 cm⁻¹ lines is sufficiently close to the 71 cm⁻¹ activation energy measured for the appearance of the new 3191.1 cm⁻¹ line for us to consider these energies to be the same, within error. Therefore, the 3265.0 and 3191.1 cm⁻¹ lines are attributed to transitions originating from two different initial states, with an energy difference of 73.9 cm⁻¹, to a common excited state (Fig. 4).

For an interatomic separation, r = 0.788 Å, consistent with a theoretically predicted [10-12] extension of ~0.04Å from the free-molecule value, one obtains a rotational energy $\hbar^2/(\mu r^2)$ of 80.7 cm⁻¹ for HD. This value is close to that observed here for the difference in frequency (73.9 cm⁻¹) of the 3265.0 and 3191.1 cm⁻¹ lines observed for HD. That this energy difference lies close to the rotational energy of a free HD molecule strongly suggests that the 3265.0 and 3191.1 cm⁻¹ lines are due to ro-vibrational transitions and that the HD molecule, while vibrating as a whole within a small interstitial region, rotates nearly freely.

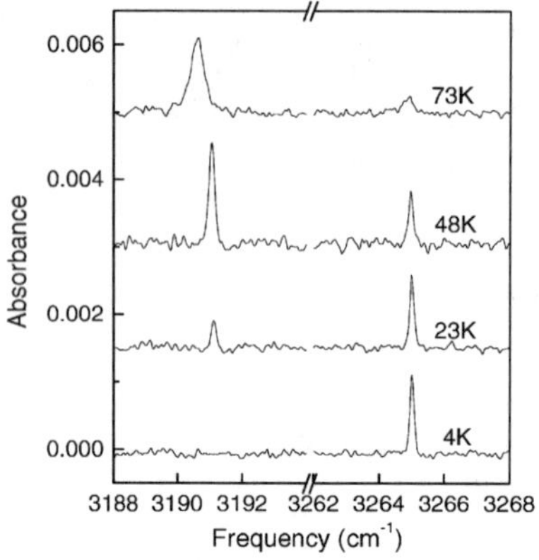
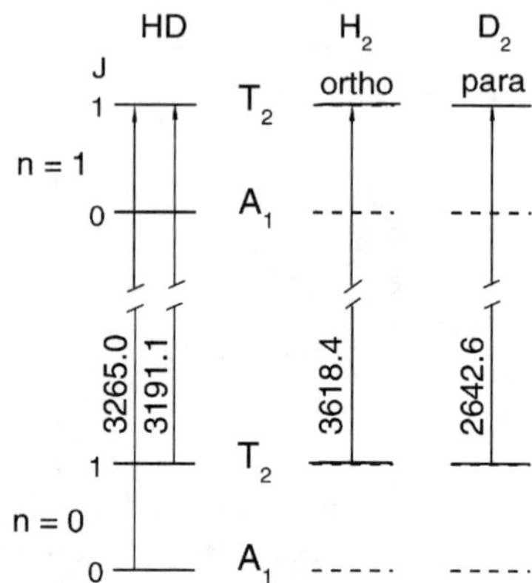

Figure 3. Spectra measured as a function of temperature for the 3265.0 cm^{-1} and 3191.1 cm^{-1} absorption lines assigned to the interstitial HD molecule in Si.

Figure 4. Energy level diagrams for the vibrational transitions of HD, ortho-H_2 and para-D_2 in Si.

The results and conclusions for HD in Si must also be consistent with the behavior of H_2 and D_2. The new absorption line for HD at 3191.1 cm^{-1} that is seen at elevated temperature lies close to the frequency position (3180 cm^{-1}) predicted from the positions of the H_2 and D_2 lines and a simple model of the anharmonicity of the hydrogen molecule [15]. Therefore, our results suggest that it is the new 3191.1 HD line and the 3618.4 and 2642.6 cm^{-1} lines of H_2 and D_2 that arise from similar transitions (see Fig. 4). Furthermore, the new 3191.1 cm^{-1} HD line is more intense than the 3265.0 cm^{-1} line, resolving the problem of the HD absorption being anomalously weak. For H_2 and D_2, the requirement that only rotational states with either odd or even parity exist depending on the parity of the nuclear-spin state makes ro-vibrational transitions with $\Delta J = 0, 2$ ... the only possibilities. For HD, there is no corresponding requirement, so all rotational states may exist. Therefore, the 3618.4, 3191.1, and 2642.6 cm^{-1} lines are assigned to purely vibrational transitions with $\Delta J = 0$. The 3265.0 cm^{-1} line is reassigned to a $J = 0$ to $J = 1$ transition that is possible only for HD. These considerations lead to the energy level diagrams for H_2, HD and D_2 shown in Fig. 4. The 3265.0 cm^{-1} line seen at 4.2K corresponds to a $J = 0$ to $J = 1$ transition, obeying the $\Delta J = \pm 1$ free rotator selection rule for a dipole-allowed transition, and thus includes one rotational quantum more than the purely vibrational transition. The 3191.1 cm^{-1} line that is observed once the $J = 1$ state is thermally populated corresponds to a $J = 1$ to $J = 1$ transition. Thus $\Delta J = 0$ and this transition corresponds to the purely vibrational transitions that must also occur for the *homonuclear* molecules H_2 and D_2 where transitions with $\Delta J = \pm 1$ are not possible.

We have noted that a dipole moment may be induced on the H_2 molecule by virtue of its zero-point motion, which dynamically breaks the exact equivalence of the two atoms [14,17]. Recently, Hourahine and Jones have pointed out that a dipole moment on the neighboring Si atoms is induced by the quadrupole moment of the rotating H_2 molecule, even when the molecule is on center [22]. They have estimated the relative contributions to the absorption

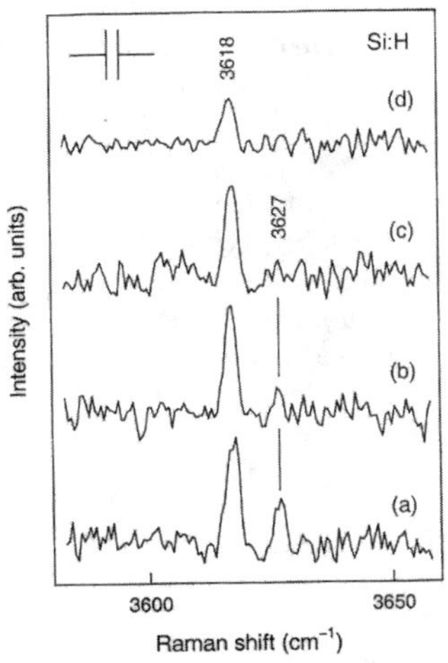

Figure 5. Raman lines for ortho-H_2 (3618 cm^{-1}) and para-H_2 (3627 cm^{-1}) in Si. Spectrum (a) was measured immediately after treatment with a hydrogen plasma. Spectra (b), (c), and (d) were measured after the sample had been stored at room temperature for one day, two days, and 18 days, respectively. (From Lavrov and Weber, ref. [20].)

strength from the induced dipole on H_2 due to its being off center, and from that induced on the neighboring silicons, and have concluded that the latter contribution dominates to a considerable degree. To test their conclusion, we have carried out density-functional-theory computations using the CRYSTAL98 code [23] and a 32-atom Si supercell, with H_2 inserted statically at various locations and directions near the T_d interstitial position. Our results depend on the choices of atomic basis sets as well as atomic locations and directions, and hence are not unambiguous; however, in general, we find that the contributions from the dipole induced on the H_2 molecule are as large as, if not larger than, the contributions induced on the neighboring silicons. It is likely that this larger H_2 dipole arises from charge density in the interstitial region, which is neglected in the estimate of Hourahine and Jones.

The $\Delta J = \pm 1$ selection rule for allowed dipole transitions of a free rotator is found to be broken for the $J = 1$ state of HD where only a J-conserving transition is seen, and not for the $J = 0$ state because the $J = 0$ to $J = 0$ transition is not observed in our spectra. To explain these selection rules, we note that in tetrahedral symmetry, $J = 0$ becomes A_1, while $J = 1$ becomes T_2. A_1 to A_1 transitions remain electric dipole forbidden in the tetrahedral field, while T_2 to T_2 transitions become electric dipole allowed (see Fig. 4). Therefore, the selection rule for rotational transitions of a free rotator, $\Delta J = \pm 1$, is broken in tetrahedral symmetry for the $J = 1$ state but not for the state with $J = 0$. If we apply these selection rules for J to H_2 and D_2, we immediately recognize that only the nuclear-spin configurations that correspond to the initial states with $J = 1$, ortho for H_2 and para for D_2, will give rise to IR absorption lines. Thus the absence of an ortho-para splitting in the IR spectra of H_2 and D_2 is explained naturally by the selection rules for J found from our ro-vibrational spectra for HD.

Surprisingly, ortho and para species were not seen in the early Raman experiments where the Raman lines of H_2 and D_2 in Si [8] were discovered. Recent Raman experiments (Fig. 5.), however, do reveal vibrational lines for both the ortho- and para-species [20]. It was also found in these experiments that para-H_2 disappears preferentially from the Raman spectrum, on the time scale of days, when the samples are stored at room temperature. This preferential disappearance of para-H_2 explains why both the ortho- and para-species were not seen in earlier

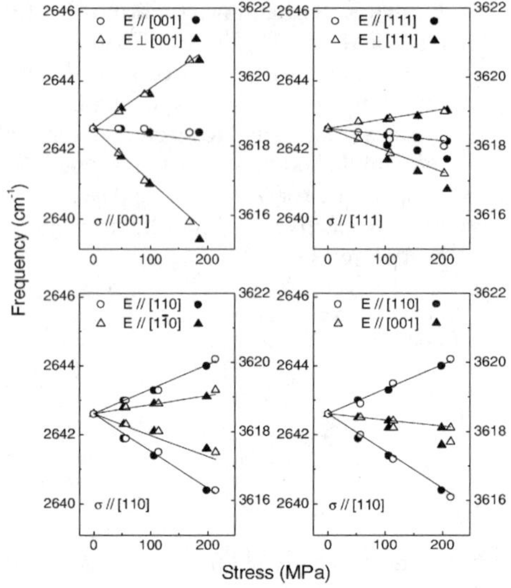

Figure 6. Stress-induced frequency shifts vs the magnitude of the stress applied along the high symmetry crystal directions for the 3618.4 cm^{-1} and 2642.6 cm^{-1} lines assigned to interstitial H_2 and D_2 in Si. The left scales correspond to data measured for D_2 (open circles and triangles) and the right scales correspond to the data measured for H_2 (filled circles and triangles).

Raman studies [8]. To explain the preferential disappearance of para-H_2, it was suggested that para-H_2 has a diffusivity that is much larger than that of ortho-H_2 and, therefore, para-H_2 diffuses out of the sample volume probed by Raman spectroscopy more quickly than ortho-H_2 [20].

CONFIRMATION OF ROTATIONAL MOTION: UNIAXIAL STRESS RESULTS

Uniaxial stress results for the H_2 and D_2 vibrational lines, previously interpreted in terms of a non-rotating defect [15], can be reinterpreted to confirm our model of a nearly free rotator [18]. The shifts and splittings of the H_2 and D_2 vibrational lines for different orientations of the applied stress are shown in Fig. 6. Here the data for H_2 and D_2 are plotted together with the frequency axes, left for D_2 and right for H_2, shifted so that the zero-stress frequencies of the D_2 and H_2 lines coincide. The 3618.4 and 2642.6 cm^{-1} transitions both show sizeable stress splittings, with the remarkable characteristic that the line splittings seen for H_2 and D_2 are nearly identical.

The static model that previously was assumed to fit the stress data for the 3618.4 cm^{-1} H_2 line [15] does not explain the isotope independence of the stress splittings seen for H_2 and D_2 and also predicts the existence of additional weak lines that are not observed. Our proposal that the IR transitions of interstitial H_2 and D_2 in Si are allowed only when the molecule is in the $J = 1$ rotational state [17] suggests a new interpretation of the stress results in which it is the $J = 1$ rotational state of the H_2 molecule that is split by stress. Kaplyanskii [24] has determined the splittings of the representations of the T_d point group for stresses applied along the high-symmetry crystal directions and also the intensities of transitions between the split states. The IR transitions seen for H_2 (and D_2) when in the $J = 1$ rotational state are T_2 to T_2 transitions. To fit our data, we have assumed that the T_2 rotational states associated with the ground and excited vibrational states show the same splittings under stress and that there is a small hydrostatic shift

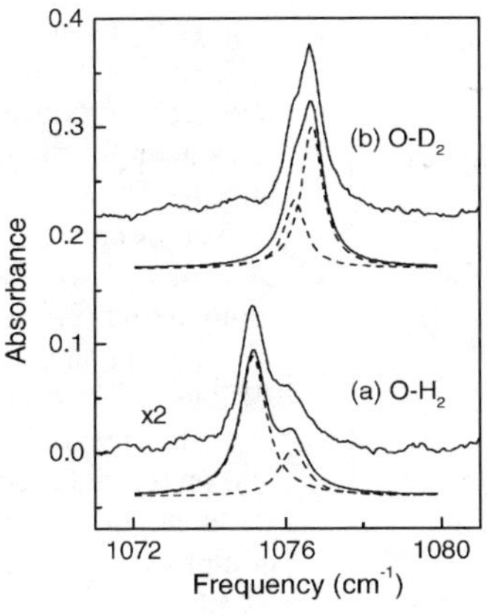

Figure 7. IR absorption spectra (4.2K) of the oxygen modes of (a) the O-H$_2$ and (b) the O-D$_2$ complexes in Si. Fits to the line shapes are shown. The samples are Cz Si into which H or D had been introduced, followed by an anneal at room temperature for at least one week.

of the center of gravity of the split lines. The straight lines drawn in Fig. 6 are the result of a least squares fit to the stress data for the 2642.6 cm^{-1} D$_2$ line. The experimental data for the positions, relative intensities, and polarization selection rules are in excellent agreement with the results expected [24] for a T$_2$ to T$_2$ transition of a defect with full tetrahedral symmetry.

The fit to our stress data for D$_2$ also provides a reasonable fit to the data for H$_2$ shown in Fig. 6. This absence of an isotope dependence, while initially puzzling, is in fact a natural result of the splitting of a degenerate rotational level, in the present case J = 1. To first order in perturbation theory, the uniaxial-stress splitting arises from a perturbing potential energy whose effect is calculated from the matrix elements of rotational wave functions. Both the potential energy and the rotational wave functions are isotope independent, so the resulting splittings are as well [18]. Model calculations by Hourahine and Jones also find stress splittings that are independent of isotope for the H$_2$ and D$_2$ molecules in Si [22].

Uniaxial stress results establish that the 3618.4 and 2642.6 cm^{-1} absorption lines seen for interstitial H$_2$ and D$_2$ in Si are due to vibrational transitions between ro-vibrational states with rotational quantum number J = 1 (T$_2$ in T$_d$ symmetry). The small dependence of the line splittings on isotope is a *signature* of the lifting of the degeneracy of the rotational states by stress.

THE O-H$_2$ COMPLEX IN Si

An O-H$_2$ complex (Fig. 1) is formed in Czochralski-grown (Cz) Si when H$_2$ migrates and becomes trapped by an O impurity [7,25,26]. O-H$_2$ gives rise to two near-lying oxygen modes (Fig. 7) and to high-frequency H$_2$ modes [7,19] that are anticorrelated with the presence of the vibrational mode of isolated H$_2$. (HD and D$_2$ in Cz Si also become trapped by O. The oxygen modes for O-D$_2$ are also shown in Fig. 7. Spectra for the high frequency modes for D$_2$ and O-D$_2$ are shown in Fig. 8.) These results were interpreted previously in terms of two O-H$_2$ complexes with different structures [7]. Our results for H$_2$ in Si [17] suggest a new model in which the H$_2$ molecule in the O-H$_2$ complex is a nearly free rotator and that both para-H$_2$ and ortho-H$_2$ are trapped by O impurities [19].

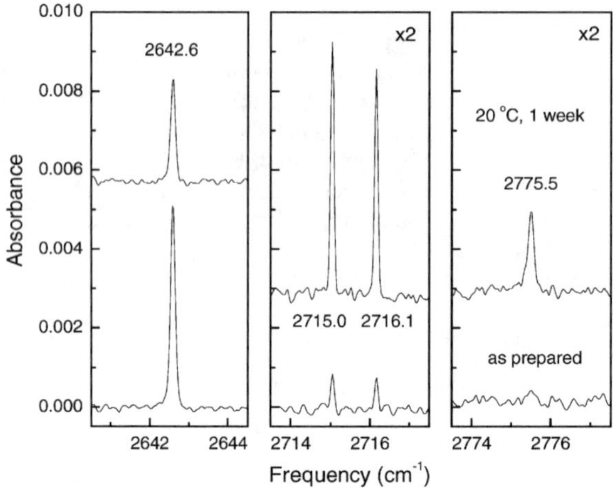

Figure 8. IR spectra (4.2K) for a Cz Si sample into which D had been introduced. Results are shown for the sample shortly after D was introduced (lower spectra) and following an anneal (1 week) at room temperature (upper spectra). The spectra on the left show the IR line for isolated interstitial D_2. The spectra in the center and on the right show the high-frequency lines for the O-D_2 complex.

In the studies of HD in Si described above, we found a new vibrational absorption line that appears at elevated temperature (T>20K). In studies of O-HD, vibrational spectra were measured as a function of temperature for Cz Si into which H and D had been introduced to search for new lines that might be associated with the O-HD complex. Fig. 9 shows spectra for the O-HD complex. As the temperature is increased above 4.2K, the 3285.4 and 3304.4 cm⁻¹ lines previously assigned to O-HD [7] become weaker while two new lines at 3282.2 and 3340.8 cm⁻¹ grow in. The joint appearance of the two new lines of O-HD at 3282.2 and 3340.8 cm⁻¹ as the temperature increases, while the two lines at 3285.4 and 3304.4 cm⁻¹ both weaken, suggests that all of these lines are due to the same defect complex. Similar to this finding for O-HD, we also conclude that the related vibrational lines of O-H_2 (and O-D_2) cannot be due to two complexes with different structures [19].

Our conclusion that only one O-H_2 structure is being observed suggests that the two near-lying oxygen lines observed for O-H_2 are due to the ortho and para states of the H_2 molecule in the complex, and similarly for O-D_2. This possibility was originally considered by Pritchard *et al.*, but was rejected in favor of an alternative model [7]. Fig. 7 shows spectra of Cz-Si samples measured after the indiffusion of (a) H and (b) D and an anneal at room temperature to produce the complexes. In spectra (a) and (b), two strong near-lying oxygen lines are seen. For the O-H_2 complex, spectrum (a) was fit by two lines with Voigt shapes and with the ratio of the intensities of the two components being 3:1. For the O-D_2 complex, spectrum (b) was again fit by two lines with Voigt shapes, but with the ratio of the intensities of the two components being 2:1. This analysis of the intensities of the oxygen modes of the O-H_2 and O-D_2 complexes is in excellent agreement with the previous results of Pritchard *et al.* [7]

The relative intensities of the two oxygen modes, 3:1 for O-H_2 and 2:1 for O-D_2, are consistent with the ratios expected for ortho to para species, where the ratios of the intensities are given by the ratios of the degeneracies of the ortho and para nuclear-spin states [23]. These results, and the knowledge that isolated, interstitial H_2 and D_2 in Si exhibit rotational motion [17, 18], lead to the conclusion that there are two near-lying oxygen lines for the O-H_2 (and O-D_2)

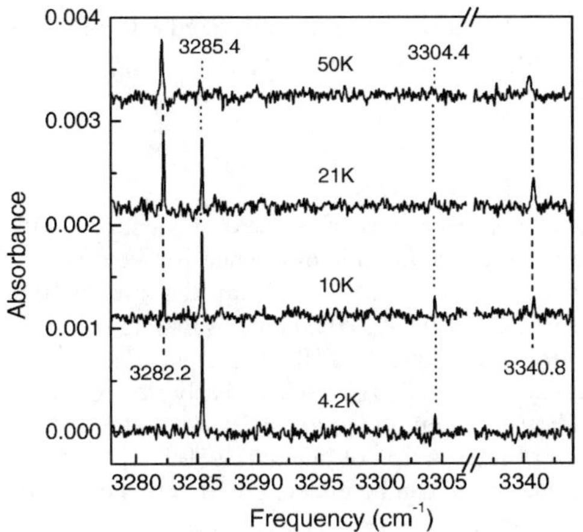

Figure 9. Spectra measured as a function of temperature for the 3282.2, 3285.4, 3304.4, and 3340.8 cm^{-1} absorption lines assigned to the O-HD complex in Si. The vibrational lines present at 4.2K are indicated with the dotted vertical lines. The lines that appear at elevated temperature are indicated with the dashed vertical lines. The sample is Cz Si into which both H and D had been introduced, followed by an anneal at room temperature for at least one week.

complex because the oxygen vibrational frequency depends on whether its H_2 (or D_2) partner is in the ortho or para state.

Our studies of the high-frequency H_2 and D_2 lines of the O-H_2 and O-D_2 complexes show that each complex has at least three high-frequency modes [19]. (See Fig. 8 for O-D_2.) These can be understood as transitions of ortho and para H_2 and D_2 trapped by oxygen. The important differences between the isolated interstitial molecules and H_2 and D_2 trapped by O are that the presence of the O atom lowers the symmetry of the neighboring interstitial site, splitting the $J = 1$ rotational states and also making H_2 (and D_2) vibrational transitions in the $J = 0$ rotational state electric-dipole allowed. The splittings of the high-frequency modes of O-H_2 and O-D_2 and the intensities of the lines can be understood in the context of this model. Energy level diagrams for the vibrational transitions of O-D_2 are shown in Fig. 10. From the spectra in Fig. 8 and the energy level diagram in Fig. 10, it can be seen that the vibrational lines assigned to O-para-D_2 (2715.0 and 2775.5 cm^{-1}) and O-ortho-D_2 (2716.1 cm^{-1}) grow in as the line due to isolated para-

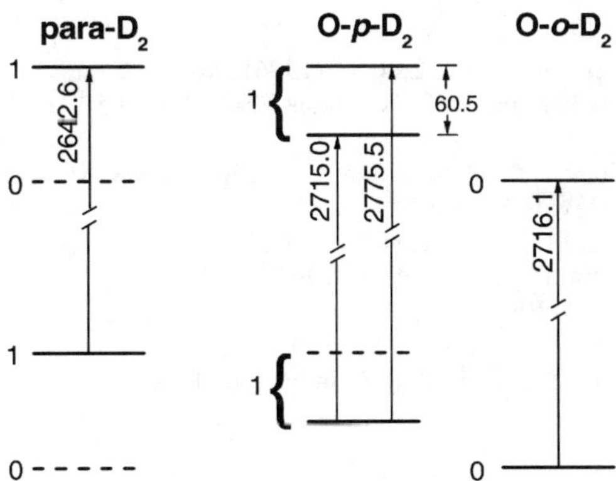

Figure 10. Energy level diagrams for the IR transitions of isolated para-D_2 in Si (left), and the complexes with oxygen, O-para-D_2 (middle) and O-ortho-D_2 (right).

D_2 (at 2642.6 cm^{-1}) disappears. (There is no IR-allowed line due to isolated ortho-D_2 in Si, whereas both para- and ortho-D_2 are seen when complexed with oxygen.)

CONCLUSION

Previous studies of interstitial H_2 in Si have led to a number of perplexing questions, with a particularly controversial issue being whether H_2 in Si is static or is a nearly free rotator. The key to the solution of this problem was the discovery of a new IR absorption line for HD in Si at 3191.1 cm^{-1} [17]. The line previously observed for HD at 3265.0 cm^{-1} is reassigned to a ro-vibrational transition that can occur only for the heteronuclear HD molecule. These results and their interpretation lead to the conclusion that interstitial H_2 in Si is a nearly free rotator. Uniaxial stress results for H_2 and D_2, which were taken previously as evidence for a static defect [15], actually provide a strong confirmation of the rotation of interstitial H_2 [18]. Furthermore, the microscopic properties of an O-H_2 complex in Si can be understood in terms of ortho and para H_2 molecules that become trapped by O impurities [19]. The vibrational lines for both ortho- and para-H_2 in Si have been seen in recent Raman experiments [20], but the preferential disappearance of para-H_2 in these experiments was unexpected and remains difficult to understand [27].

ACKNOWLEDGMENTS

We thank G. D. Watkins, F. S. Ham (deceased), J. Weber, and E. V. Lavrov for numerous helpful discussions. This work was supported by NSF Grant No. DMR-0108914, and NREL Grant No. AAT-2-31605-4.

REFERENCES

[1] A. Mainwood and A. M. Stoneham, Physica 116B (1983) 101.
[2] J. W. Corbett, S. N. Sahu, T. S. Shi, and L. C. Snyder, Phys. Lett. 93A (1983) 303.
[3] S. K. Estreicher, Mat. Sci. Eng. R 14 (1995) 319.
[4] R. N. Hall, J. Elect. Mat. 14a (1985) 759.
[5] J. Vetterhöffer, J. Wagner and J. Weber, Phys. Rev. Lett. 77 (1996) 5409.
[6] R. E. Pritchard, M. J. Ashwin, J. H. Tucker, and R. C. Newman, Phys. Rev. B 57 (1998) 15048.
[7] R. E. Pritchard, M. J. Ashwin, J. H. Tucker, R. C. Newman, E. C. Lightowlers, M. J. Binns, S. A. McQuaid, and R. Falster, Phys. Rev. B 56 (1997) 13118.
[8] A. W. R. Leitch, V. Alex, and J. Weber, Phys. Rev. Lett. 81 (1998) 421.
[9] Y. Okamoto, M. Saito, and A. Oshiyama, Phys. Rev. B 56 (1997) 10016.
[10] C. G. Van de Walle, Phys. Rev. Lett. 80 (1998) 2177.
[11] C. G. Van de Walle and J. P. Goss, Mater. Sci. Eng. B 58 (1999) 17.
[12] B. Hourahine, R. Jones, S. Öberg, R. C. Newman, P. R. Briddon, and E. Roduner, Phys. Rev. B 57 (1998) 12666.

[13] S. K. Estreicher, K. Wells, P. A. Fedders and P. Ordejón, J. Phys.: Condens. Matter 13 (2001) 6271.

[14] W. B. Fowler, P. Walters and M. Stavola, Phys. Rev. B 66 (2002) 075216.

[15] J. A. Zhou and M. Stavola, Phys. Rev. Lett. 83, (1999) 1351; J. A. Zhou, E Chen, and M. Stavola, Phys. Rev. Lett. 84 (2000) 4778.

[16] R. C. Newman, R. E. Pritchard, J. H. Tucker, and E. C. Lightowlers, Phys. Rev. B 60 (1999) 12775

[17] E E. Chen, M. Stavola M, W. B. Fowler, and P. Walters, Phys. Rev. Lett. 88 (2002) 105507.

[18] E E. Chen, M. Stavola M, W. B. Fowler and J. A. Zhou, Phys. Rev. Lett. 88 (2002) 245503.

[19] E E. Chen, M. Stavola M, and W. B. Fowler, Phys. Rev. B 65 (2002) 245208.

[20] E. V. Lavrov and J. Weber, Phys. Rev. Lett. 89 (2002) 215501.

[21] R. Eisberg and R. Resnick, Quantum Physics of Atoms, Molecules, Solids, Nuclei and Particles, 2nd ed., New York, Wiley, 1985.

[22] B. Hourahine and R. Jones, Phys. Rev. B 67 (2003) 121205.

[23] V. R. Saunders, R. Dovesi, C. Roetti, M. Causá, N. M. Harrison, R. Orlando, and C. M. Zicovich-Wilson, CRYSTAL 98 User's Manual, Univ. Torino, Torino, Italy, 1998.

[24] A. A. Kaplyanskii, Opt. Spectrosc. (USSR) 16 (1964) 557.

[25] V. P. Markevich, M. Suezawa, and K. Sumino, Mater. Sci. Forum 196-201 (1995) 915.

[26] V. P. Markevich and M. Suezawa, J. Appl. Phys. 83 (1998) 2988.

[27] M. Stavola, E E. Chen, W. B. Fowler, and G. A. Shi, Physica B 340-342 (2003) 58.

Comparative Study of Electronically Controlled Motion of Hydrogen around Carbon and Platinum Atoms in Silicon

Yoichi Kamiura, Namula Bao, Kimihiro Sato, Kazuhisa Fukuda[1], Yasuyuki Iwagami, Yoshifumi Yamashita and Takeshi Ishiyama
Faculty of Engineering, Okayama University, 3-1-1 Tsushima-naka,
Okayama 700-8530, Japan
[1]NEC Laboratories, Otsu 520-0833, Japan

ABSTRACT

We have studied the local motion of hydrogen in the neighborhood of carbon and platinum impurities by observing the stress-induced reorientation and subsequent recovery of two H-related (H-C and Pt-H_2) complexes in Si, using deep-level transient spectroscopy (DLTS) under uniaxial compressive stress. We notice two interesting differences in hydrogen motion around carbon and platinum atoms. The first one is a difference in the temperature where stress-induced reorientation occurs. That of the H-C complex occurs at high temperatures above 250 K, while it occurs at low temperatures around 80 K for the Pt-H_2 complex. The second difference is the effect of charge state of the complexes on their stress-induced reorientation and subsequent recovery. It occurs preferentially when an electron occupies the level of the H-C complex, but the Pt-H_2 complex has the reverse effect of level occupancy. These differences are discussed from viewpoint of different atomic configurations and electronic states of two H-related complexes.

INTRODUCTION

The static properties of hydrogen in semiconductors, such as electronic state, stable lattice location and configuration, have extensively been studied, but only little information has been obtained on the dynamic properties of hydrogen. Such information as diffusion and local motion around defects and impurities is useful to design electronic devices considering the spatial extent where hydrogen influences the electrical and optical properties of semiconductors. However, hydrogen is so mobile and reactive to other defects and impurities that one cannot detect hydrogen in isolated states but mostly in bound states to other defects and impurities.

Few groups have studied the local motion of hydrogen bound to impurities. The dynamical tunneling motion of proton was observed in some systems, H-Si in Ge [1] and H-Be in Si [2]. On the other hand, the hydrogen motion around boron in Si is not pure tunneling but thermally assisted tunneling [3]. Recently, we have found the signatures of electronically controlled motion of hydrogen bound to carbon in Si [4, 5], and observed that hydrogen moves faster in the electron-occupied charge state than in the electron-empty one in the recovery of stress-induced reorientation of the H-C complex. The same charge-state-dependent motion of isolated hydrogen was observed in the recovery of stress-induced reorientation of the AA9 EPR center [6]. More recently, we have observed the completely opposite charge-state-dependence of hydrogen motion around platinum in the Pt-H_2 complex in Si [7].

In this paper, we compare electronically controlled motion of hydrogen in the neighborhood of two impurities (carbon and platinum) in Si, and show the unique effects of electronic states of hydrogen and its surroundings on the local motion of hydrogen, which was directly probed by measuring the recovery of stress-induced reorientation of the above two H-related defect complexes by means of DLTS under uniaxial stress.   We discuss differences in charge-state effects from viewpoint of different atomic configurations and electronic states of the complexes.

EXPERIMENTAL PROCEDURE

Two FZ grown n-type silicon crystals doped with phosphorus were used.   One crystal contains the carbon impurity with a density of 1×10^{16} cm^{-3} and was used to study the H-C complex.   The other one is carbon-free and was used to study the Pt-H$_2$ complex.   Crystals were cut into square pillars with dimensions of $1 \times 1 \times 6$ mm^3, the longest of which was parallel to the <100>, <110> or <111> direction.   Platinum was introduced into samples by evaporation and subsequent annealing at 850°C for 2 h in an argon ambience.   Hydrogen or deuterium was injected by chemical etching (HF : HNO$_3$ (DNO$_3$)= 1 : 10).   Schottky contacts were formed by vacuum evaporation of gold.   Uniaxial compressive stresses up to 1.2 GPa were applied to the samples along their longest dimension during capacitance DLTS measurements and annealing experiments for stress-induced reorientation.   We applied reverse bias voltage to the Schottky junction to control the charge state of H-C and Pt-H$_2$ complexes during the annealing. Other details were the same as published previously [4, 5, 8, 9].

HYDROGEN MOTION AROUND CARBON

Figure 1 shows a structural model of the H-C complex proposed by our group.   Hydrogen occupies the bond-centered (BC) site between carbon and silicon atoms.   Under an <110> stress, the H-C complex is reoriented to the configuration with the lowest total energy.   We observed that this lowest-energy configuration corresponded to the configuration where hydrogen occupied the bonds perpendicular to the stress direction.   This means that hydrogen jumps from a high-energy bond to a low-energy one in the stress-induced reorientation of the H-C complex under an <110> stress, as indicated in figure 1.   Subsequently, if stress is removed, stress-induced alignment of the configuration is lost by random jumps of hydrogen among bonds.   We studied such local motion of hydrogen in the neighborhood of carbon by observing the recovery kinetics of stress-induced reorientation of the H-C complex [10].   The results are shown in figure 2, where jump rates for hydrogen in the electron-occupied and electron-empty charge

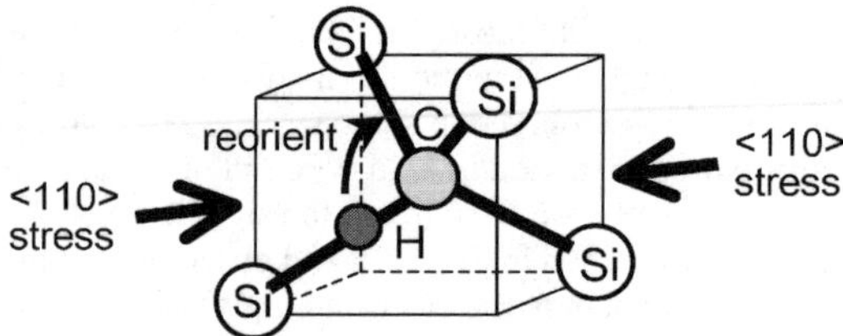

Figure 1. Structural model of the H-C complex and hydrogen motion in the stress-induced reorientation of the H-C complex under an <110> stress.

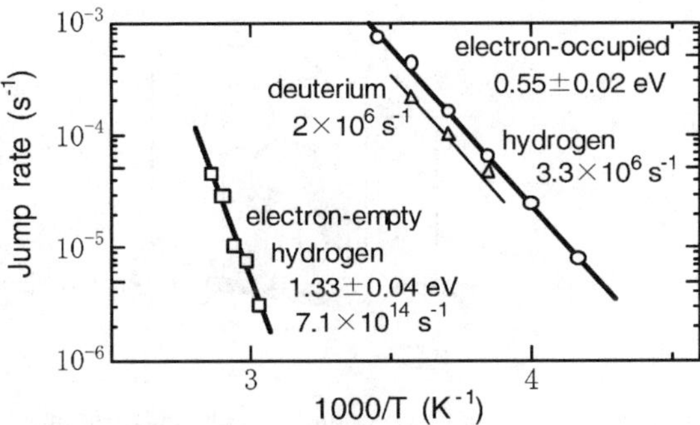

Figure 2. Temperature dependence of jump rates of hydrogen and deuterium around carbon, evaluated from the recovery process of stress-induced reorientation of H-C and D-C complexes.

states and that for deuterium in the electron-occupied charge state are represented by circles, squares and triangles, respectively. The activation energies and frequency factors determined by the fitting for hydrogen motion are indicated beside thick solid lines. A thin solid line represents a fitting with an activation energy fixed to 0.55 eV for deuterium motion, giving a smaller frequency factor by a factor of 0.6 than that for hydrogen motion, indicating that this recovery process is actually due to hydrogen jumps. A remarkable feature is that the activation energy and frequency factor for hydrogen motion is greatly reduced by changing the charge state of the H-C complex from electron-empty to electron-occupied states. As for the frequency factor, the magnitude (7.1×10^{14} s^{-1}) in the electron-empty state is consistent with the usual atomic jump frequency, while that in the electron-occupied state is smaller and strongly suggests that hydrogen motion is limited by the electron capture from the conduction band. The change in activation energy will be discussed later.

HYDROGEN MOTION AROUND PLATINUM

We studied the local motion of hydrogen in the neighborhood of platinum by observing the kinetics of stress-induced reorientation of the Pt-H$_2$ complex [7, 11, 12]. Hydrogen was much more mobile in this complex than in the H-C complex, so that the stress-induced reorientation occurred even during the DLTS temperature scan. Accordingly, we applied isothermal DLTS (IT-DLTS) technique with no temperature scan [13], instead of conventional DLTS, at low temperatures of 70 K and below, where hydrogen could not move. We determined the symmetry of the Pt-H$_2$ complex to be orthorhombic with the C$_{2v}$ point group from an observed splitting pattern of IT-DLTS peaks under stress in three major directions. Based on this results, we proposed a structural model of the Pt-H$_2$ complex, as shown in figure 3, where two hydrogen atoms are directly bonded to the platinum atom. There are two sets of two silicon atoms that are displaced inward. The Pt-H$_2$ entity is located between the two sets. This model has a symmetrical axis along the <100> direction with C$_{2v}$ symmetry.

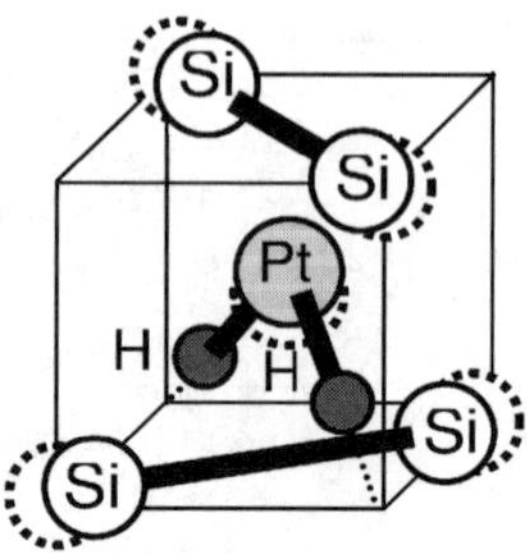

Figure 3. Structural model of the Pt-H_2 complex deduced from an observed splitting pattern of IT-DLTS peaks under stresses applied along three major directions.

We observed that the Pt-H_2 complex was reoriented at temperatures as low as 80 K under a uniaxial stress of 0.6 GPa applied along the <111> direction and a reverse bias of 3V. Accordingly, site changes of hydrogen are expected to be much faster in the Pt-H_2 complex than in the H-C complex.   We performed a series of isothermal annealing experiments to obtain quantitative results of stress-induced reorientation of the Pt-H_2 complex.   We analyzed the data and obtained the temperature dependence of jump rates of hydrogen and deuterium around platinum.   The result is shown in figure 4, where the activation energy of hydrogen jumps is only 0.27 eV, much lower than that of hydrogen motion around carbon.   Isotope effect is confirmed as well.   We have recently studied the recovery process of stress-induced reorientation, and have found that the activation energy of hydrogen jumps is also very low, about 0.25 eV.   These results suggest that stress has little effect on hydrogen motion.

The stress-induced reorientation occurred only when an electron did not occupy the level of the Pt-H_2 complex, indicating a clear charge-state effect on the local motion of hydrogen around the Pt atom, similarly to the H-C complex.   However, the effect of level occupancy was totally opposite for both complexes.   Under the above bias condition, most of Pt-H_2 complexes were in

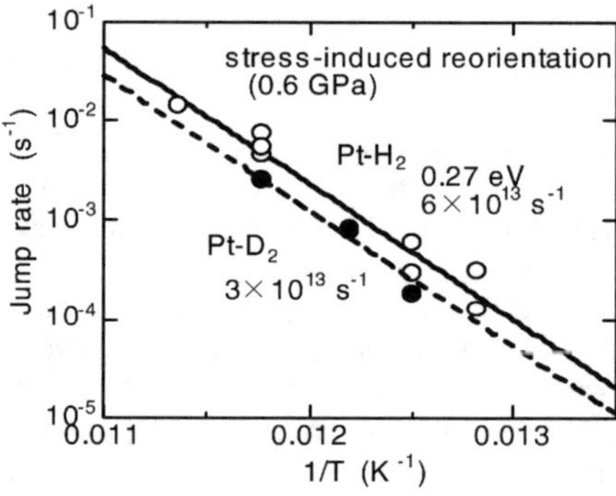

Figure 4. Temperature dependence of jump rates of hydrogen and deuterium around platinum, evaluated from the stress-induced reorientation of Pt-H_2 and Pt-D_2 complexes.

the singly negative charge state at temperatures around 80 K. We observed no changes of IT-DLTS spectra without applying reverse bias [11], indicating that hydrogen was mobile only in the singly negative charge state of the complex. We have recently observed the same charge-state effect on hydrogen motion in the recovery process of stress-induced reorientation. Such differences in the effect of level occupancy will be discussed in the following section.

DISCUSSION

If we compare hydrogen motion around the carbon atom with that around the platinum atom, we notice two interesting differences. The first one is a difference in the temperature where stress-induced reorientation occurs. That of the H-C complex occurs at high temperatures above 250 K, while it occurs at temperatures as low as 80 K for the $Pt-H_2$ complex. The second difference is the effect of charge state of the complexes on their stress-induced reorientation and recovery. It occurs preferentially when an electron occupies the level of the H-C complex, but the $Pt-H_2$ complex has the reverse effect of level occupancy. The activation energy for the hydrogen motion is 1.33 eV and 0.55 eV in the electron-empty and electron-occupied charge states of the H-C complex, respectively, and 0.25 eV in the electron-empty charge states of the $Pt-H_2$ complex.

These differences may result from different atomic configurations and electronic states of two H-related complexes. In the H-C complex, hydrogen is located at the BC site between carbon and silicon atoms. The electronic state of the complex lying in the band gap has antibonding character [14]. In the electron-empty charge state, the magnitude of a frequency factor (7.1×10^{14} s^{-1}) is consistent with the usual atomic jump frequency. The activation energy, 1.33 eV, is much higher than those (0.43-0.44 eV) of both local motion [6] and long-range diffusion [15] of isolated hydrogen in Si. The difference, 0.9 eV, may arise from the binding energy between hydrogen and carbon. In other words, there is a higher energy barrier for hydrogen motion in the C-H-Si system than in the Si-H-Si one. If the H-C complex captures an electron from the conduction band at its gap state with antibonding character, the complex raises its electronic energy by the energy difference between bonding and antibonding states, typically 1 eV and more. On the other hand, in both of C-H-Si and Si-H-Si systems, hydrogen may feel similar potentials at the saddle points between adjacent BC sites, because hydrogen is located apart from C and Si atoms. Therefore, the electron capture considerably lowers the energy barrier for hydrogen motion in the electron-occupied charge state. The captured electron lies in the gap state with the antibonding character, which makes the C and Si atoms relaxed outward to lower the energies of the C-H-Si system. When hydrogen reaches the saddle point, all atoms around it are appropriately relaxed to lower the energy of the saddle-point configuration, further reducing the energy barrier. In this way, hydrogen motion is limited by the electron capture. In other words, hydrogen waits for the events of electron capture to jump more quickly.

In the $Pt-H_2$ complex, the two hydrogen atoms are directly bonded to the platinum atom, and therefore defect reorientation needs no bond switching but only the rotation of the whole $Pt-H_2$ entity. This may well fit with the present results that the reorientation occurs very easily with a low activation energy. Here, it is reasonable to assume that the electronic level arises from the electronic state whose wave function is distributed among the platinum atom and two hydrogen neighbors along the <110> axis. Our previous results of the stress effect on the electronic state strongly suggests that compressive stress to the wave function raises its energy, indicating its antibonding character [12].   If an electron occupies such electronic state with antibonding character, the two hydrogen atoms may be displaced outward, probably retarding their motion for the reorientation.   An essential difference between H-C and Pt-H_2 complexes may be that the energy difference between bonding and antibonding states more strongly affects the total energy of the defect system for the H-C complex.

CONCLUSIONS

Comparing the local motion of hydrogen in the neighborhood of carbon with that of platinum in Si, we notice two essential differences.   The first one is a difference in the temperature where stress-induced reorientation occurs.   The second difference is the effect of charge state of the complexes on their stress-induced reorientation and subsequent recovery. These differences are discussed from viewpoint of different atomic configurations and electronic states of H-C and Pt-H_2 complexes, and are ascribed to that the energy difference between bonding and antibonding states more strongly affects the total energy of the defect system for the H-C complex.

ACKNOWLEDGMENT

This work was supported in part by a Grant-in-Aid for Scientific Research (No.15340099) from the Ministry of Education, Culture, Sports, Science and Technology.

REFERENCES

1.   E. E. Haller, *Phys. Rev. Lett.* **40**, 584 (1978).
2.   K. Muro and A. J. Sievers, *Phys. Rev. Lett.* **57**, 897 (1986).
3.   Y. M. Cheng and M. Stavola, *Phys. Rev. Lett.* **73**, 3419 (1994).
4.   Y. Kamiura, N. Ishiga and Y. Yamashita, *Jpn. J. Appl. Phys.* **36**, L1419 (1997).
5.   K. Fukuda, Y. Kamiura and Y. Yamashita, *Physica B* **273-274**, 184 (1999).
6.   Y. V. Gorelkinskii, in *Hydrogen in Semiconductors II*, edited by N. H. Nickel (Academic, 1999), p. 25.
7.   K. Fukuda, Y. Iwagami, Y. Kamiura, Y. Yamashita and T. Ishiyama, *Physica.* **B308-310**, 240 (2001).
8.   M. Yoneta, Y. Kamiura and F. Hashimoto, *J. Appl. Phys.* **70**, 1295 (1991).
9.   Y. Kamiura, M. Hayashi, Y. Nishiyama, S. Ohyama and Y. Yamashita, *Jpn. J. Appl. Phys.* **36**, 6579 (1997).
10. Y. Kamiura, K. Fukuda, Y. Yamashita and T. Ishiyama, *Phys. Rev.* **B65**, 113205 (2002).
11. Y. Kamiura, Y. Iwagami, K. Fukuda, Y. Yamashita, T. Ishiyama and Y. Tokuda, *Microelectronic Engineering* **66**, 352 (2003).
12. Y. Kamiura, K. Sato, Y. Iwagami, Y. Yamashita, T. Ishiyama and Y. Tokuda, *Phys. Rev.* **B69**, 045206 (2004).
13. Y. Tokuda and K. Kamiya and T. Okumura, *J. Appl. Phys.* **88**, 1943 (2000).
14. K. Fukuda, Y. Kamiura, Y. Yamashita and T. Ishiyama, *Jpn. J. Appl. Phys.* **40**, 6700 (2001).
15. B. Holm, K. Bonde Nielsen and B. Beck Nielsen, *Phys. Rev. Lett.* **66**, 2360 (1991).

Vibrational Lifetimes of Hydrogen in Silicon: Isotope Effects and MOSFET Reliability

Gunter Lüpke[1], Baozhou Sun[1], Norman H. Tolk[2], and Leonard C. Feldman[2]

[1]Department of Applied Science, The College of William and Mary, Williamsburg, Virginia 23187

[2]Department of Physics and Astronomy, Vanderbilt University, Nashville, Tennessee 37235

ABSTRACT

Characterization of defect and impurity reactions, dissociation and migration in semiconductors requires detailed understanding of rates and pathways of vibrational energy flow, of energy transfer channels and of coupling mechanisms between local modes and the phonon bath of the host material. Significant progress in reaching this goal has been accomplished in recent landmark studies exploring the excitation and dynamics of vibrational states associated with hydrogen in silicon. We describe recent experiments which measure the vibrational lifetime of the Si-H bond in various defect configurations and show the relationship between these lifetimes and silicon MOSFET reliability.

INTRODUCTION

Recently, the dynamics of hydrogen-related defects in silicon has attracted much attention [1]. Because of its small size, hydrogen (H) is easily incorporated in silicon during the growth process and can form different defects. Figure 1 shows the two-dimensional structure of hydrogen defects in silicon. The major role of hydrogen is the passivation of both shallow and deep impurity centers and other defects through the formation of complexes [2]. On the surface, H forms covalent bonds with Si dangling bonds [3]. In the bulk, H forms all kinds of interstitial, self-interstitial, vacancy-type defects or complexes with other impurities [4]. At the interface, H passivation can reduce the density of P_b centers, which is crucial in semiconductor devices [5]. These H-related defects have been well characterized [6,7].

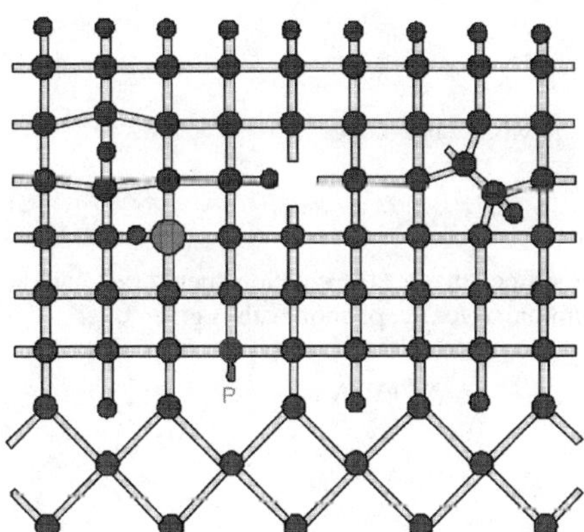

Figure 1. Two-dimensional model of crystalline silicon (blue dots) with various hydrogen-decorated complexes (red dots).

The addition of impurities to a host crystal, whose masses are less than the masses of the host atoms, gives rise to new vibrational modes with higher frequencies than the host crystal's vibrational (phonon) modes [8]. The vibrational spectroscopy of these high-frequency modes has become an important probe of defects in solids and often provides information about defect structure and properties that cannot be obtained by other methods [9]. However, relatively little is known about the dynamics of local vibrational modes (LVMs) in condensed matter systems in spite of the importance of the coupling of LVMs to the phonon bath of the host material [10,11]. Coupling between LVMs and the heat bath is responsible for the flow of energy into and out of impurity and defect complexes. This is important, because when defects act as recombination centers in semiconductors, the electronic excitation that is deposited at the defects is then available to promote defect migration and reactions, an area of defect physics that has an important impact on the degradation of electronic and optoelectronic devices. Hence a deeper understanding of the physical properties of defects and impurities in semiconductors can be obtained by studying the dynamics of the elastic and inelastic local interactions.

THEORETICAL BACKGROUND

The level of theoretical investigation into the lifetime of local vibrational modes is less advanced. This state of affairs is partly because of the difficulties in explicitly treating slow relaxation processes in complex systems, and partly because, as suggested by experiment, a highly anharmonic coupling mechanism is apparently responsible for the relaxation process. Expressions for the vibrational relaxation and dephasing rates have been derived by defining an effective vibrational Hamiltonian that represents canonical averages over the bath modes [12,13]. By expanding the system-bath-interaction Hamiltonian to high-order in the phonon coordinates an expression for the multi-phonon relaxation rate is derived.

If the vibrational relaxation process involves transferring energy directly to the bath modes (phonons) in a single step, and the amount of energy transferred exceeds by many times the maximum energy of a single phonon, then many phonons will have to be created simultaneously. This process is known as multiphonon relaxation. One might reasonably imagine that the rate of such a high-order process would be quite small. Therefore it is remarkable that even if the number of phonons emitted is as high as ten or more, multiphonon processes typically compete successfully with radiative decay, and are in fact often the dominant relaxation mechanism.

The vibrational Hamiltonian can be written as

$$H = H_S + H_B + H_I ,\qquad (1)$$

where the Hamiltonian of the local vibrational mode,

$$H_S = \hbar\omega a^\dagger a ,\qquad (2)$$

is characterized in terms of the frequency ω, the creation operator $a^\dagger$, and the annihilation operator a. The (harmonic) Hamiltonian for the phonon bath is given by

$$H_B = \sum_k \hbar\omega_k b_k^\dagger b_k ,\qquad (3)$$

where $b_k^\dagger$ and b_k are the phonon creation and annihilation operators, respectively. There are at least three different possible routes to multi-phonon relaxation based on the form of the interaction Hamiltonian H_I [13]. In the most common approach H_I is strictly off-diagonal in the two-level basis of the local vibrational mode [12,13]. The off-diagonal matrix elements of H_I are expanded to high order in the phonon coordinates, and are treated by lowest-order time-dependent perturbation theory. The interaction Hamiltonian is then given by

$$H_I = \sum_{\{v\}} \hbar \left(G_{\{v\}} B_{\{v\}} a^\dagger + G_{\{v\}}^* B_{\{v\}}^\dagger a \right), \tag{4}$$

where

$$B_{\{v\}} = \prod_{i=1}^{N_v} b_{i,\{v\}} \; , \; B_{\{v\}}^\dagger = \prod_{i=1}^{N_v} b_{i,\{v\}}^\dagger \; , \tag{5}$$

and $G_{\{v\}}$ denotes the temperature-independent coupling strength of the channel $\{v\}$, which is characterized by the set $\{\omega_1^{(v)}, \omega_2^{(v)}, \ldots, \omega_{N_v}^{(v)}\}$ of accepting mode frequencies. Energy is conserved in the decay process,

$$\hbar\omega = \sum_{j=1}^{N_v} \hbar\omega_j^{(v)} \; . \tag{6}$$

The contribution to the decay rate due to H_I is given by the standard result of first-order time-dependent perturbation theory. The total decay rate (inverse lifetime) is given as the sum of the rates of all of the decay channels [12,13]:

$$\frac{1}{T_1} = 2\pi \sum_{\{v\}} |G_{\{v\}}|^2 \, n_{\{v\}} \rho_{\{v\}} \; , \tag{7}$$

The function $n_{\{v\}}$ describes the temperature-dependent population of the receiving modes:

$$n_{\{v\}} = \frac{\exp(\beta\hbar\omega) - 1}{\prod_{j=1}^{N_v} \left[\exp(\beta\hbar\omega_j^{(v)}) - 1 \right]} \; . \tag{8}$$

The compound spectral density of accepting states, $\rho_{\{v\}}$, can be expressed in terms of a convolution of single spectral densities of states

$$\rho_{\{v\}} = \int d\omega_1^{(v)} \ldots \int d\omega_{(N_v-1)}^{(v)} \, \rho_{(1)}^{(v)}(\omega_{(1)}^{(v)}) \ldots \rho_{N_v}^{(v)}(\omega_{N_v}^{(v)}) \; . \tag{9}$$

In the low-temperature limit, the decay rate reflects spontaneous decay into N_v accepting modes and $n_{\{v\}} \cong 1$. At higher temperatures the decay rate increases due to stimulated emission in a fashion determined by the frequencies of the accepting modes [12,13]. As a rule, the process with the least amount of phonons should be favored.

EXPERIMENTAL DETAILS

Lifetimes ranging from a few picoseconds to several nanoseconds can be measured directly by the pump-probe transient bleaching technique utilizing ultrashort pulses from a tunable infrared laser. Typically, two different tunable, infrared lasers are used: ultrafast Optical Parametric Amplifiers (OPAs) or picosecond Free-Electron Lasers (FELs). These lasers can deliver pulses with a time duration of ~1 ps, spectral width of ~15 cm^{-1}, and pulse energies of several µJ. Both lasers can in principle be tuned in wavelength between 1-10 µm, thus covering the Si-H stretch vibrations. The major difference between these lasers is the pulse repetition frequency (PRF), which for OPAs is typically limited to a few kHz, whereas FELs can achieve much higher PRFs. For example, the FEL at the Thomas Jefferson National Accelerator Facility (TJNAF) operates at a PRF of 18 MHz [14]. A high PRF and excellent pulse-to-pulse stability are necessary to achieve a high signal-to-noise ratio in pump-probe measurements.

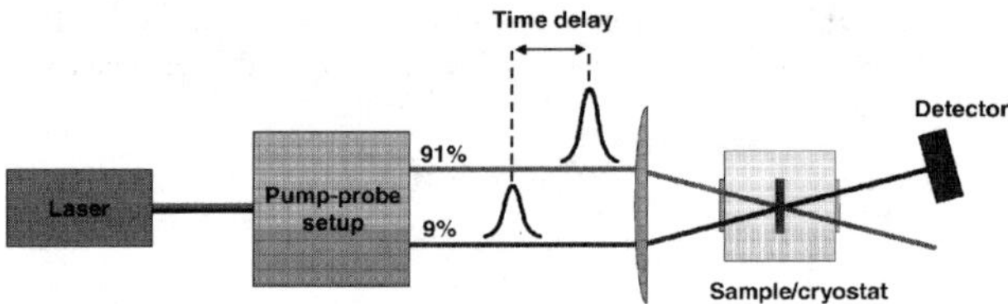

Figure 2. Schematic of the pump-probe transient bleaching experiment.

A schematic diagram of a typical setup for a pump-probe transient bleaching experiment is shown in figure 2. The laser beam is split into two parts, pump and probe, carrying 91% and 9% of the power, respectively. The two incident ultrashort laser pulses are spatially overlapped on the sample. The pump excites a fraction of the Si-H bonds to the first excited state, which causes a transient increase in the transmission coefficient of the sample that decays over time due to the decay of the excited mode. The time evolution of the transient bleaching signal,

$$S_b \propto \exp[-\tau/T_1], \tag{10}$$

is monitored by varying the time delay (τ) between the pump and probe pulses. The signal S_b is detected with an amplified liquid nitrogen-cooled HgCdTe or InSb detector. To improve the sensitivity of the experiment a lock-in amplifier is utilized along with an optical chopper which modulates the excitation for phase-sensitive detection.

DISCUSSION

Recently, we undertook a systematic study of the vibrational lifetime of various H and deuterium (D) related point defects in bulk silicon and germanium [15-18]. The low-temperature lifetime of H at the bond-center site in crystalline Si and other interstitial-like complexes was found to be of the order of 10 ps (figure 3), while the lifetime of vacancy-H complexes are 1 – 2 orders of magnitude greater. The observed large disparity in measured lifetimes is unexpected based on simple theories and is as yet unexplained. These initial results indicate that local defect structure plays a crucial role in the coupling mechanism.

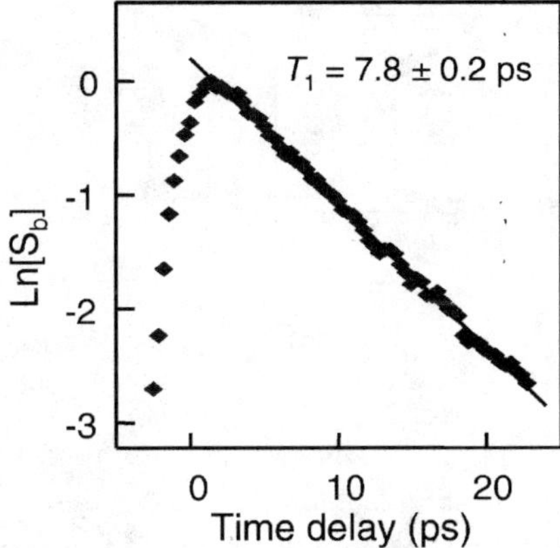

Figure 3. Decay of the transient bleaching signal from the 1998-cm^{-1} line of bond-center hydrogen in silicon measured at ~20 K with the TJNAF FEL [15].

The strong structure dependence might be explained by either pseudo-localized modes or LVMs being involved in the decay process. Being localized around these defects, such modes have much larger vibrational amplitudes on the atoms close to the Si-H bond than lattice phonons, which may cause a strongly enhanced anharmonic coupling to phonons of the Si-H stretch mode. However, significant distortions of the Si-Si bonds in the vicinity of the defects are required for pseudo-localized modes or LVMs to form, which is consistent with highly distorted interstitial-type defects having shorter lifetimes than vacancy defects. For the vacancy defects, it certainly seems plausible that the relatively large open volume of the vacancies would lead to a small interaction between the hydrogen and the surrounding silicon atoms, which would lead to a longer vibrational lifetime.

Figure 4 shows the structure of various hydrogen-decorated defects with their lifetimes. The vibrational lifetimes of H_{BC}, H_2^*, $HV \cdot VH_{(110)}$ are measured directly by transient bleaching technique described above. Some are estimated by their absorption line width through the equation

$$\Gamma_0 = \frac{1}{2\pi cT_1} + \frac{1}{\pi cT_2^*} + \Gamma_{inhom},\qquad(11)$$

where Γ_0 is the linewidth of absorption line, T_1 is the lifetime of the defect, T_2^* is the dephasing time and Γ_{inhom} is the inhomogeneous linewidth. The first term determines the natural linewidth of the mode, whereas the second term generally dominates the linewidth at elevated temperatures due to elastic scattering of the local mode with phonons. A good lower estimate of the vibrational lifetime can be obtained from infrared absorption spectra when the measurements are done at low temperature on a sample with a low concentration of defects and impurities.

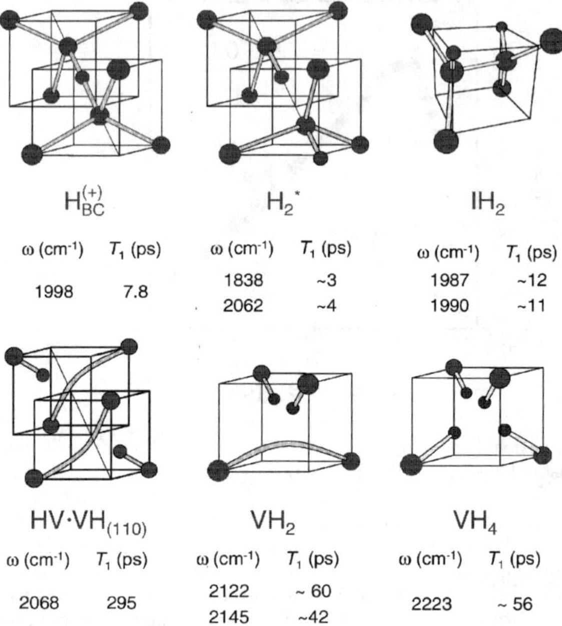

Figure 4. Structures of selected H-related defects in crystalline silicon. The top (bottom) row shows interstitial-type (vacancy-type) defects. The blue spheres are Si atoms and the red spheres are H atoms. The vibrational frequencies and lifetimes are also listed.

From our studies of Si-H stretch modes, we know that the vibrational dynamics of H defects is determined by anharmonic coupling with modes within the phonon bands. Two or three low-frequency modes are required to obtain qualitative agreement with the temperature dependence of the lifetimes. The nature of these exchange modes is presently not known. However, likely candidates are perturbed acoustic or optical phonons, or Si-H bend modes.

Deuterium lifetimes and the reliability of Si MOSFETS

Table I shows the lifetimes for a selection of stretch modes of H and D-related defects in crystalline silicon measured at 10 K [16]. Substitution of H with D affects the lifetime of LVM's in two ways. First, the vibrational amplitude is less for Si-D modes than for Si-H modes, and since the decay of the LVM into other modes is an anharmonic effect, Si-D modes have longer lifetimes than the corresponding Si-H modes. Second, the vibrational energy of the Si-D mode is $\sqrt{2}$ less than that of the Si-H mode. Consequently, the order of the decay process, i.e., the number of modes into which the LVM decays, can be smaller for Si-D than for Si-H. Since the strengths of anharmonic terms are expected to decrease with increasing order, this suggests that Si-D modes have shorter lifetimes than the corresponding Si-H modes. Until recently, the second contribution has been assumed to be dominant [19], which is consistent with the experimental results for Si-H and Si-D modes on H-terminated Si surfaces [20]. In contrast, table I. shows that the lifetimes of D related stretch modes are longer than the corresponding Si-H lifetimes for all defects except HV·VH$_{(110)}$, for which the Si-D lifetime is 3 times shorter than for Si-H.

Table I. Vibrational lifetimes for a selection of stretch modes of H and D-related defects in crystalline silicon measured at 10 K [16].

Defect	ω_H (cm^{-1})	T_1 (ps)	T_1 (ps)	ω_D(cm^{-1})
H_2^*	2062.1	1.9	4.8	1500.1
IH_2	1987.1	12	20	1446.5
IH_2	1990.0	11	18	1448.7
VH_2	2122.3	60	70	1547.9
VH_2	2145.1	42	55	1565.1
VH_4	2223.0	56	143	1617.5
HV·VH$_{(110)}$	2072.5	295	93	1510.4

Lyding et al. demonstrated that H plays an active role in the degradation process of silicon MOSFET's by showing that devices treated with deuterium (D) have transconductance lifetimes 10-50 times longer than H-passivated devices [21]. In general, the degradation effect is understood as the electron impact induced desorption of hydrogen (deuterium), which excites the bond to either a dissociative electronic state or to excited vibrational states. Unpassivated, these dangling bonds represent sufficient scattering centers to seriously degrade performance. This "giant isotope effect" was in agreement with earlier results of Shen at al [22] and Foley et al [23] who showed that the probability of desorption of deuterium on clean silicon surfaces was significantly smaller than for hydrogen desorption.

Based on well-understood surface science experiments the bond strength of both Si-H and Si-D is ~3eV. The explanation for these isotope effects then requires an understanding of how low energy electrons (<3eV) can induce desorption of hydrogen and why deuterium desorption rates are smaller than hydrogen desorption rates. This was explained most simply by Persson et al [24] in their truncated harmonic oscillator model (sometimes referred to as the thermal heating model). In this picture there is a competition between excitation to higher levels of the oscillator and the decay to lower levels. With a sufficiently high excitation rate, compared to the decay rate, it is possible to work up the energy levels. If a sufficiently high level is achieved desorption may occur. For example in the hydrogen-silicon system the first harmonic level corresponds to ~0.25eV excitation energy. Thus excitation to the 12th level is required to reach the 3eV value of the binding energy.

Using this model Persson et al derived a general expression for the dissociation rate, R, associated with excitation to the nth channel as

$$R \sim \frac{n+1}{T_1}\left[\frac{\Gamma_{exc}}{\Gamma_{exc}+T_1^{-1}}\right]^{n+1}, \tag{12}$$

where Γ_{exc} is the excitation rate, due to electron collisions, and T_1^{-1} is the decay rate, similar to the lifetime measurements described above. Unfortunately, it is not possible to measure the actual lifetime of passivating hydrogen in a genuine MOSFET structure. There is insufficient hydrogen for such an experiment. However we can attempt to evaluate equation (12), using estimates of the excitation rate and the measured lifetime in a configuration as close to the dangling bond as possible.

For the lifetime we use the value of 295 ps for the HV-VH center shown in table I. Alternatively we might use the value of ~800 ps measured by Guyot-Sionnest et al [20] for the bare surface. In any case the decay rate is clearly in this range. For the excitation probability we resort to atomic physics concepts. The excitation probability is given as

$$\Gamma_{exc} = \sigma \cdot j, \qquad (13)$$

where σ, corresponds to the cross section for ~0.25 eV excitation by a low energy electron and j is the lateral current density in the MOSFET. The current density in a typical MOSFET is estimated as 6×10^{26} electrons/cm^2/sec, corresponding to a MOSFET current of 1mA and an effective area of $(1\mu m) \times (10\text{Å})$. The atomic physics literature [25] indicates that the cross-section for excitation is $\sim 10^{-16}$/cm^2, so that $\Gamma_{exc} = 6 \times 10^8$/sec where we have assumed that all electrons within the channel are available for excitation. Thus the excitation rates and the decay rates are comparable and multiple excitation through the harmonic levels is indeed possible. Using a standard MOSFET configuration Chen et al [26] confirmed this general picture by showing that the degradation rate is dependent on the current and not the integrated charge.

It is interesting to note that R is always less than 1. For deuterium desorption one must reach the 17th level since the harmonic-like energy level is reduced by $\sqrt{2}$ and the binding energy of 3 eV remains the same as for hydrogen. Then it is clear that since R< 1 the desorption rate for deuterium is significantly less than for hydrogen, even accounting for some increase in the excitation cross-section.

CONCLUSION

In summary, we have studied the lifetimes of a selection of Si-H stretch modes of point defects in Si. The lifetimes of interstitial-type defects were found to be very short (1.6-8 ps), whereas vacancy-type defects have lifetimes up to several hundreds of picoseconds – two orders of magnitude variation. The strong dependence of the lifetime on the atomic structure of the defects suggests that pseudo-localized modes or LVMs are involved in the vibrational relaxation of Si-H bonds of point defects in solids. We have also discussed how these lifetime measurements give an indication of the fundamental parameters controlling Si MOSFET reliability values.

ACKNOWLEDGMENTS

The authors thank Michael Budde for fruitful discussions and preparation of hydrogen-implanted Si samples. We gratefully acknowledge the staff at the TJNAF FEL and W. M. Augustyniak for technical assistance. This work was supported in part by DOE through Grant No. DE-FG02-99ER45781 (C.W.M and V.U.), ONR (C.W.M and V.U.), NSF through Grants No. DMR-00-76027/DMR-02-42316 (C.W.M), and Tomas F. and Kate Miller Jeffress Memorial Trust through Grant No. J-545 (C.W.M).

REFERENCES

1. G. Lüpke, N. H. Tolk and L. C. Feldman, J. Appl. Phys. **93**, 2316 (2003).
2. J. I. Pankove and N. M. Johnson, 1991, *Hydrogen in Semiconductors* (Semiconductors and Semimetals **34**) (New York: Academic).
3. G. S. Higashi, Y. J. Chabal, G. W. Trucks, and K. Raghavachari, Appl. Phys. Lett. **56**, 656 (1990).
4. *Hydrogen in Semiconductors II:* Semiconductors and Semimetals, edited by N. H. Nickel (Academic Press, New York, 1999), Vol. **61**, and references therein.
5. A. Stesmans, Phys. Rev. B **61**, 8393 (2000); J. Appl. Phys. **88**, 489 (2000).
6. *Identification of Defects in Semiconductors: Semiconductors and Semimetals*, edited by M. Stavola (Academic Press, New York, 1999), Vol. **51B**, Chap. 3, and references therein.
7. S. J. Pearton, J. W. Corbett, and M. Stavola, Hydrogen in Crystalline Semiconductors (Springer Verlag, Berlin, 1992), and references therein.
8. A. S. Barker, Jr., and A. J. Sievers, Rev. Mod. Phys. **47**, Suppl. No. 2, S1 (1975).
9. "Vibrational Spectroscopy of Light Element Impurities in Semiconductors," M. Stavola, in *Identification of Defects in Semiconductors*, vol. **51b**, edited by M. Stavola (Academic Press, New York, 1998), chapter 3, p. 157; and references therein.
10. A. Laubereau and W. Kaiser, Rev. Mod. Phys. **50**, 607 (1978).
11. D. D. Dlott, in *Laser Spectroscopy of Solids* II, ed. by W. M. Yen, Topics in Applied Physics, vol. **65** (Springer-Verlag, Berlin, 1989), p. 167.
12. A. Nitzan and J. Jortner, Mol. Phys. **25**, 713 (1973); A. Nitzan, S. Mukamel, and J. Jortner, J. Chem. Phys. **60**, 3929 (1974).
13. S. A. Egorov and J. L. Skinner, J. Chem. Phys. **103**, 1533 (1995).
14. G. R. Neil, C. L. Bohn, S. V. Benson, G. Biallas, D. Douglas, H. F. Dylla, R. Evans, J. Fugitt, A. Grippo, J. Gubeli, R. Hill, K. Jordan, R. Li, L. Merminga, P. Piot, J. Preble, M. Shinn, T. Siggins, R. Walker, and B. Yunn, Phys. Rev. Lett. **84**, 662 (2000).
15. M. Budde, G. Lüpke, C. Parks Cheney, N. H. Tolk, L. C. Feldman, Phys. Rev. Lett. **85**, 1452 (2000).
16. M. Budde, G. Lüpke, E. Chen, X. Zhang, N. H. Tolk, L. C. Feldman, E. Tarhan, A. K. Ramdas, M. Stavola, Phys. Rev. Lett. **87**, 145501 (2001).
17. G. Lüpke, X. Zhang, B. Sun, A. Fraser, N. H. Tolk, L. C. Feldman, Phys. Rev. Lett. **88**, 135501 (2002).
18. C. Parks Cheney, M. Budde, G. Lüpke, N. H. Tolk, L. C. Feldman, Phys. Rev. B **65**, 35214 (2002).
19. C. G. Van De Walle, and W. B. Jackson, Appl. Phys. Lett. **69**, 2441 (1996).
20. P. Guyot-Sionnest, P. Dumas, Y. J. Chabal, and G. S. Higashi, Phys. Rev. Lett. **64**, 2156 (1990)
21. J. W. Lyding, K. Hess, and I. C. Kizilyalli, Appl. Phys. Lett. **68**, 2526 (1996).
22. T. C. Shen, C. Wang, G. C. Abeln, J. R. Tucker, J. W. Lyding, P. Avouris, R. E. Walkup, Science **268**, 1590 (1995).
23. E. T. Foley, A. F. Kam, J. W. Lyding, P. Avouris, Phys. Rev. Lett. **80**, 1336 (1998)
24. B. N. J. Persson, P. Avouris, Surface Science **390**, 45 (1997).
25. G. J. Schulz, Rev. Mod. Phys. **45**, 423 (1973).
26. Z. Chen, P. Ong, A. K. Mylin, V. Singh, and S. Chetlur, App. Phys. Lett. **81**, 3278 (2002).

Origin of the Hydrogen/Deuterium (H/D) Isotope Effect of Hot-Electron Degradation of MOS Devices

Zhi Chen, Jun Guo, and Pangleen Ong
Department of Electrical and Computer Engineering and Center for Micro-Magnetic and
Electronic Devices, University of Kentucky, Lexington, KY 40506, U.S.A.

ABSTRACT

In order to verify Van de Walle and Jackson's theory on the isotope effect of the Si-H/D bonds resistant to hot-electron excitation [*Appl. Phys. Lett.*, **69**, 2441 (1996)], we measured the Si-H, Si-D, and other vibrational modes in oxidized silicon wafers annealed in hydrogen and deuterium using Fourier Transform Infrared (FTIR) spectrometry. Our FTIR data suggest that the frequency for the Si-D bending mode at the SiO_2/Si interface is 490 cm^{-1}. Our experimental data support Van de Walle and Jackson's theory with some modification. Their theory is correct for the experiments of breaking Si-H/D bonds using scanning tunneling microscope (STM) where no oxide is involved. In the SiO_2/Si case, the de-excitation of the Si-D bond may be due to the energy coupling from the Si-D bending mode to two vibrational modes; i.e., the Si-O TO mode and the Si-Si TO phonon mode. Van de Walle and Jackson only pointed out coupling to the Si-Si TO phonon mode. The strongest coupling might happen between the Si-D mode and the Si-O TO mode. Therefore, the oxide may play a crucial role in energy dissipation of the Si-D bond in metal-oxide-semiconductor (MOS) devices.

INTRODUCTION

The giant hydrogen/deuterium (H/D) isotope effect was discovered in the study of the desorption of hydrogen (H) and deuterium (D) on silicon in ultra high vacuum (UHV) using Scanning Tunneling Microscopy (STM) [1,2]. This effect was later used in passivation of the SiO_2/Si interface, leading to large improvement of the hot-electron-related lifetime of metal-oxide-semiconductor (MOS) transistors using deuterium process [3-9].

Van de Walle [10,11] proposed a theory to explain this remarkable isotope effect, stating that the Si-D bond is more resistant to hot-electron excitation than the Si-H bond. The Si-H/D bond-breaking at the SiO_2/Si interface is caused by two competing processes. One is that the energy of the bonds is accumulated through excitation by energetic hot electrons. The other process is de-excitation where the bond energy is taken away by coupling between the Si-H/D vibrational modes and substrate phonons. Van de Walle [10] suggested that the vibrational frequency of Si-D bending mode is close to the Si-Si TO phonon mode (460 cm^{-1}), resulting in energy coupling between the Si-D bending mode and the Si-Si TO phonon mode. This de-excitation efficiently strengthens the Si-D bond. On the other hand, because the vibrational frequency of the Si-H bond is far away from the Si-Si TO phonon mode, there is no energy coupling between the Si-H bending mode and the Si-Si TO phonon mode, leading to Si-H bonds more vulnerable to hot-electron excitation.

However, no experimental data exist regarding the vibrational frequency of the Si-D bond at the SiO_2/Si interface to support the above theory. Although the vibrational frequency (510 cm^{-1}) of the Si-D bond is reported for deuterated amorphous Si (α-Si)[12], the chemical environment in the oxidized single crystal Si (SiO_2/Si structure) is very different from that in α-Si. In order to understand fundamental mechanisms of isotope effect of hot-electron degradation of MOS transistors, it is of great importance to measure the vibrational frequency of the Si-D bond and other chemical bonds in the SiO_2/Si structure and study the energy coupling from the Si/H(D) bonds to other vibrational modes. In this paper, we present the first experimental measurement of vibrational modes at the SiO_2/Si interface by Fourier Transform Infrared (FTIR) spectrometry and study the correlation between the vibrational modes and energy dissipation of the Si-D bond.

EXPERIMENTAL DETAILS

The n$^+$ silicon wafers with (100) orientation, 50 mm (2 inch) in diameter, and ~0.3 mm in thickness were used in our experiments. Because high-density electrons are present at the SiO_2/Si interface in MOS transistors, we use n$^+$ silicon wafers. The wafers were prepared using conventional RCA cleaning. The thermal oxidation was performed in dry O_2 at a constant temperature of 1050 °C (1323 K) for 15 minutes. The thickness of oxide layer was measured to be 32 nm. Some of the oxidized wafers were annealed at 450°C in 100% H_2 and 100% D_2 for 60 min separately. Because it is unlikely to have the same oxide for each wafer, in order to have accurate comparison, the oxidized wafer was cut into two half-wafers. One half-wafer was annealed in H_2 or D_2 and the other half was used for comparison. Infrared absorbance spectra of the processed wafers were obtained immediately by a Fourier Transform Infrared (FTIR) Spectrometer (Thermo Nicolet Nexus 470), in conjunction with a Variable Angle Specular Reflectance Accessory (Pike VeeMax II). The spectral range was selected from 400 to 900 cm^{-1}, with a resolution of 8 cm^{-1}. All spectra were baseline corrected after proper correction for CO_2 peaks from the air by using the OMNIC software.

RESULTS AND DISCUSSION

Fig. 1 shows the FTIR spectra of a SiO_2/Si sample with oxide thickness of ~32 nm before and after the hydrogen anneal. The two curves (before and after hydrogen anneal) are very similar. The peak (592 cm^{-1}) appears uniquely in the H-annealed sample, which is determined to be Si-H bending mode. There are five other peaks appearing in both curves with a slight difference in magnitude, which might be ascribed to the slight structure change due to the thermal process in annealing. The main peak at 468 cm^{-1} may be ascribed to the Si-Si TO phonon mode as suggested by Van de Walle [10,11,13] and Lin [14]. The wavenumber of 806 cm^{-1} is Si-O TO bending mode [15,16]. The peak at 505 cm^{-1} is Si-O LO mode [16] and at 448 cm^{-1} is Si-O TO rocking mode [15,16]. The peak at 435 cm^{-1} is suggested to be Si-Si LO mode [18]. Table I summarizes the vibrational modes of the as-oxidized, H-annealed, and D-annealed SiO_2/Si samples.

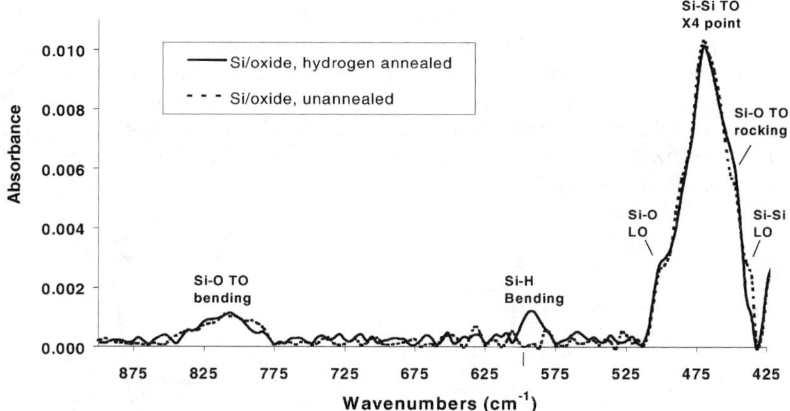

Figure 1. FTIR spectra of Si/SiO$_2$ wafers before (-----) and after hydrogen annealing (—), oxide thickness 32 nm, spectral resolution 8 cm^{-1}, 128 scans, 65 degree grazing angle.

Table I. Vibrational modes obtained by the Fourier transform infrared (FTIR) measurement

Wave Number (cm^{-1})	806	592	505	490	468	448	435
Type of Vibrational Modes	Si-O TO bending[a]	Si-H Bending	Si-O LO[b]	Si-D Bending	Si-Si TO [c]	Si-O TO Rocking[d]	Si-Si LO[e]
In as-oxidized Si	Yes	No	Yes	No	Yes	Yes	Yes
In H-annealed SiO$_2$/Si	Yes	Yes	Yes	No	Yes	Yes	Yes
In D-annealed SiO$_2$/Si	Yes	No	Yes	Yes	Yes	Yes	Yes

[a]References 15,16,
[b]Reference 16
[c]References 10,11,13
[d]References 15,16
[e]Reference 18

Fig. 2 shows the FTIR spectra of a SiO$_2$/Si sample with oxide thickness of ~32 nm before and after deuterium anneal. The peak at 490 cm^{-1} uniquely appears in the D-annealed SiO$_2$/Si sample, which should be the Si-D bending mode. The other five peaks are almost the same as those in the as-oxidized samples. It should be noted that there is no experimental data available for Si-H and Si-D vibrational modes in the oxidized *single crystal Si* samples. The existing data of the Si-H and Si-D vibrational modes were from the *amorphous Si* samples [12,17]. There is a striking difference between our data from the single crystal Si and the data from amorphous Si [12,17]. At first, the wave number of the Si-Si TO phonon mode in our sample is ~468 cm^{-1}, which is very close the value (463 cm^{-1}) obtained by other techniques [13]. However, the vibrational mode of the Si-Si TO phonon mode in the amorphous Si sample is 495 cm^{-1} as reported by Wei et al. [12]. There is a difference of 27 cm^{-1} between the single crystal Si sample and the amorphous Si sample. We believe that this is due to the structure difference between the single crystal Si and amorphous Si. Secondly, it is more interesting to note that the vibrational frequency of Si-D bond (490 cm^{-1}) in our oxidized single crystal sample (See Fig. 2) is 20 cm^{-1} lower than that (510 cm^{-1}) in the amorphous Si [12,17]. Finally, the Si-H vibrational mode (592 cm^{-1}) in our oxidized single crystal sample (See Fig. 1) is ~50 cm^{-1} lower than that (640 cm^{-1}) in the amorphous Si sample [12,17].

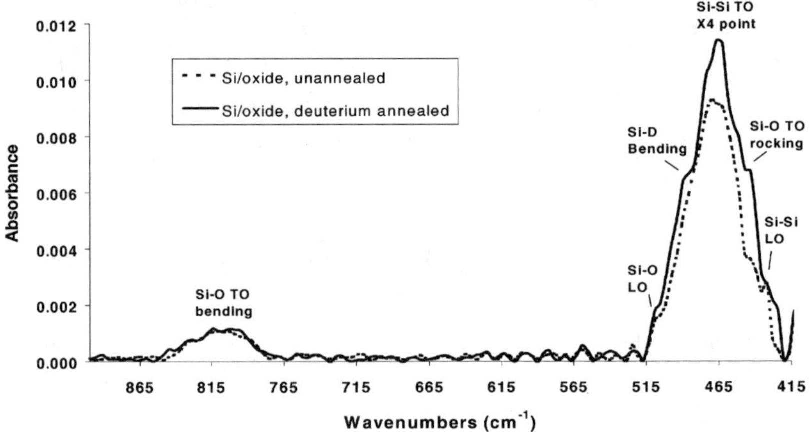

Figure 2. FTIR spectra of Si/SiO$_2$ wafers before (-----) and after deuterium annealing (——), oxide thickness 32 nm, spectral resolution 8 cm^{-1}, 128 scans, 65 degree grazing angle.

Of the most importance, it is found in Fig. 2 that the absorbance of the Si-Si TO phonon mode and the Si-O TO rocking mode are all enhanced significantly (>25%) after deuterium anneal. There is not much difference for absorbance of the Si-Si TO mode and Si-O TO rocking mode after hydrogen anneal (See Fig. 1). This exclusively suggests that there is energy coupling from the Si-D vibrational mode to the Si-Si TO phonon mode and the Si-O TO rocking mode. This proves that Van de Walle's theory is correct for experiments of STM-induced breaking of Si-H/D bonds where no oxide involves [1,2]. For the MOS transistor case, there may be two channels for de-excitation of Si-D bonds: energy coupling to the Si-O TO mode and to the Si-Si TO mode. The strongest coupling may happen between the Si-D mode and the Si-O TO mode rather than Si-Si TO phonon mode because more enhancement of absorbance of the Si-O TO mode (~60%) than the Si-Si TO phonon mode (~25%) is found in the deuterium-annealed samples (See Fig. 2).

CONCLUSIONS

In summary, our FTIR data support the basic point of Van de Walle and Jackson's theory with some modification. Van de Walle and Jackson's theory is correct for the experiments of STM-induced breaking of Si-H/D bonds where no oxide involves. For the MOS transistor case, the de-excitation of the Si-D bond may be due to the energy coupling between the Si-D mode and two vibrational modes, i.e. Si-O TO mode and the Si-Si TO phonon mode. Van de Walle and Jackson only suggested the coupling to the Si-Si TO phonon mode. The strongest coupling might happen between the Si-D mode and the Si-O TO mode. Therefore, the oxide may play a crucial role in energy dissipation of the Si-D bond in MOS devices.

ACKNOWLEDGEMENTS

This research is supported by National Science Foundation ECS-0093156 and the Office of Vice President for Research, University of Kentucky.

REFERENCES

1. J. W. Lyding, *Proc. of IEEE* **85**, 589 (1997).
2. P. Avouris, R. E. Walkup, A. R. Rossi, T.-C. Shen, G. C. Abeln, J. R. Tucker, and J. W. Lyding, *Chem. Phys. Lett.,* **257**, 148 (1996).
3. J. W. Lyding, K. Hess, and I. C. Kizilyalli, *Appl. Phys. Lett.* **68**, 2526 (1996).
4. I. C. Kizilyalli, G. Abeln, Z. Chen, G. Webber, B. Kozias, S. Chetlur, J. W. Lyding, and K. Hess, *IEEE Electron. Dev. Lett.* **19**, 444 (1998).
5. R. A. B. Devine, J.-L. Autran, W. L. Warren, K. L. Vanheusdan, and J.-C. Roastaing, *Appl. Phys. Lett.* **70**, 2999 (1997).
6. P. J. Chen and R. M. Wallace, *Appl. Phys. Lett.* **73**, 3441 (1998).
7. Z. Chen, J. Lee, J. W. Lyding, and K. Hess, *Digest of Technical Papers of IEEE Symposium on VLSI Technology*, 180 (1998).
8. Z. Chen, *Ph.D. Dissertation*, University of Illinois at Urbana-Champaign, Aug. 1999.
9. Z. Chen, K. Hess, J. Lee, J. W. Lyding, E. Rosenbaum, I Kizilyalli, S. Chetlur, and R. Huang, *IEEE Electron Dev. Lett.* **21**, 24 (2000).
10. C. G. Van de Walle and W. B. Jackson, *Appl. Phys. Lett.* **69**, 2441 (1996).
11. C. G. Van de Walle, *IEEE Trans. Electron Dev.* **47**, 1779 (2000).
12. J.-H. Wei, M.-S. Sun, and S.-C. Lee, *Appl. Phys. Lett.* **71**, 1498 (1997).
13. *Data in Science and Technology: Semiconductors*, edited by O. Madelung, Springer-Verlag, Berlin, 1991.
14. S.-Y. Lin, *J. Appl. Phys.* **82**, 5976 (1997).
15. G. Lucovsky, M. J. Manitini, J. K. Srivastava, and E. A. Irene, *J. Vac. Sci. Technol.* **B5**, 530 (1987)
16. M. K. Gunde, *Physica B* **292**, 286 (2000).
17. M. H. Brodsky, M. Cardona, and J. J. Cuomo, *Phys. Rev. B* **16**, 3556 (1977).
18. K. Kim, *J. Vac. Sci. Technol. A* **16**, 2272 (1998).

Donor Behaviour of Implanted Hydrogen Ions in Silicon Wafers

Damien Barakel and Santo Martinuzzi
TECSEN- UMR 6122 - University of Aix-Marseilles III - 13397 Marseilles – France

ABSTRACT

P type Czochralski (Cz) and float-zone (FZ) grown silicon wafers were investigated with doping levels of 5×10^{14} and 2×10^{15} cm^{-3}, respectively. Hydrogen ions are implanted at a dose of 2×10^{16} cm^{-2}, at energies in the range 20 to 250 keV and are accumulated at depth R_p in the range 0.2 to 2.4 µm. After implantation the wafers are annealed between 350 and 600°C for 30 min, under argon flow. It is found that a graduated n-p junction is formed, and after metallization a photovoltaic device is obtained, which works like a solar cell. SIMS analysis shows that, around R_p, hydrogen concentration achieves 10^{21} cm^{-3}. I-V and C-V curves confirm the formation of a N-type layer in which the donor concentration is about 5×10^{17} cm^{-3}. When the samples are annealed at temperatures higher than 550°C the counter-doping vanishes. The observed behaviour of hydrogen is irrespective of oxygen concentrations in the wafers as it occurs in Cz (oxygen rich) like in FZ (oxygen poor) wafers. If the wafers are ion implanted with helium at the same dose and energy no junction appears. It is concluded that the agglomeration of hydrogen in silicon after ion implantation at a dose exceeding 10^{16} cm^{-2} gives rise to the formation of shallow donors.

INTRODUCTION

Hydrogen has long been recognized as having an important role in modifying the electrical properties of semiconductors. In p-type silicon hydrogen is in a positive charge state (H^+), it is a fast diffuser and its motion can be influenced by internal or external electric fields [1]. In hydrogen ion implanted silicon, it should be emphasized that hydrogen can exist in two distinct states : one of these is strongly bonded to passivated dangling bonds, while the second, more mobile, could be weakly bonded interstitial atoms, accumulated below the free surface in concentration higher than the solubility limit, up to 10^{20} cm^{-3} [2]. Such interstitial hydrogen can behave as a donor [3-6], with 26 meV ionisation energy, in proton implanted silicon wafers at doses higher than 10^{16} cm^{-2}
It was demonstrated that implanted hydrogen can convert p type oxygen rich silicon in n-type by the enhanced formation of thermal donors after interaction with interstitial oxygen atoms [6]. It is also well known that implantation of H^+ ions at doses higher than 10^{16} cm^{-2} leads to the formation of gas bubbles at the projected range R_p. These bubbles are transformed into nanocavities when the gas out-diffuses during annealing at temperature higher than 550°C.

In the present paper, we confirm that H^+ implantations at high doses create n-type regions close to the free surface and form n-p junctions in p-type silicon crystals which behave like solar cells.

EXPERIMENTAL

Monocrystalline silicon samples, float-zone grown (FZ) and Czochralski grown (Cz), experienced the implantations. They were P-type boron doped to 4×10^{13} cm^{-3} (FZ samples) and 3×10^{15} cm^{-3} (Cz samples), respectively. The samples were implanted with H$^+$ and He$^+$ ions, at energies in the range 20 to 250 keV and at a dose of 3×10^{16} cm^{-2}. SIMS analysis and electronic microscopy were used to obtain the hydrogen profile and an image of the gas accumulation zone. Mesa structures (2x2 mm to 20x20 mm) were obtained by photolithography and chemical etching. Ohmic contacts resulted of the deposition of aluminium layers followed by an annealing at 350°C for 30 min. In order to make solar cells a grid is realized by the lift-off technique on the front surface. C-V as well as I-V measurements were also done to determine the parameters of the junction and the doping level of the n-type region and this level is also controlled by a four-point probe resistivity tool. The larger mesa diodes were investigated as photovoltaic devices and compared to standard solar cells (without antireflection coating) made by phosphorus in-diffusion from POCl$_3$ at 850°C for 30 min.

RESULTS

H$^+$ implanted samples exhibit a photovoltaic behaviour after the formation of ohmic contacts. Fig. 1 shows the photocurrent spectral responses obtained with a FZ sample implanted at 250 keV and annealed at various temperatures between 350 and 600°C for 1 h. The photocurrent slightly increases after each annealing and becomes comparable to that of a classical FZ solar cell in the red and near infra-red part. However when the temperature exceeds 550°C the photocurrent decreases neatly, due to hydrogen exodiffusion.
A similar behaviour is observed in Cz implanted samples, however the photocurrent does not change after the first annealing needed to form the ohmic contacts.
There is no photovoltaic behaviour in Helium implanted samples under the same conditions.
For both FZ and Cz samples the photocurrent is low between 400 and 700 nm suggesting that there is a dead zone, around 2 µm deep, between the surface and the active N-type region. Such a depth corresponds to the projected range R_p of H$^+$ ions at 250 keV given by TRIM simulation. SIMS profiles confirms the value of R_p: they detect a concentration of 10^{21} cm^{-3} atoms of hydrogen at 2.4 µm.
Thinning the dead zone by chemical etching increases the current between 400 and 700 nm, as shown by Fig. 2, and the spectral variations of the photocurrent become close to that of a standard FZ solar cell (the thinning is obtained by immersion of the samples in a CP4 solution for few seconds). When more than 2 µm are etched, the photocurrent strongly decreases. These results suggest that the n-p junction is located around R_p.
This suggestion is confirmed by the investigation of wafers implanted with the same hydrogen ion dose but at various ion energies : 20, 40 and 120 keV, as shown by Fig. 3. Notice that for both the lower ion energies the photovoltaic effect tends to disappear after

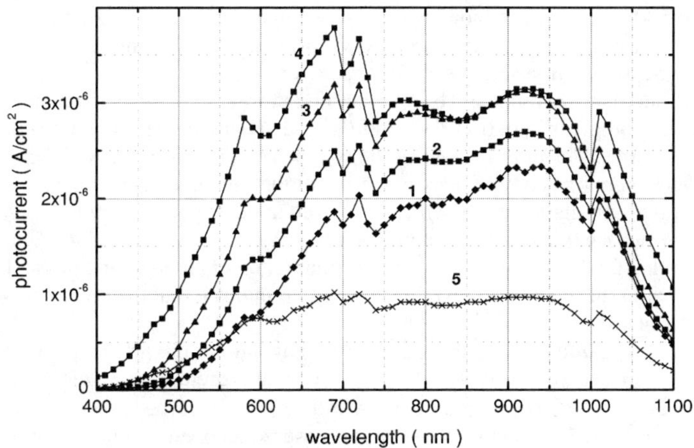

Figure 1. Spectral variations of the photocurrent of FZ samples implanted at 250 keV and annealed at 350°C for 30 min (1), at 450°C (2), 550°C (3), and 600°C (5) for 1 hour. Curve 4 is that of a classical FZ solar cell.

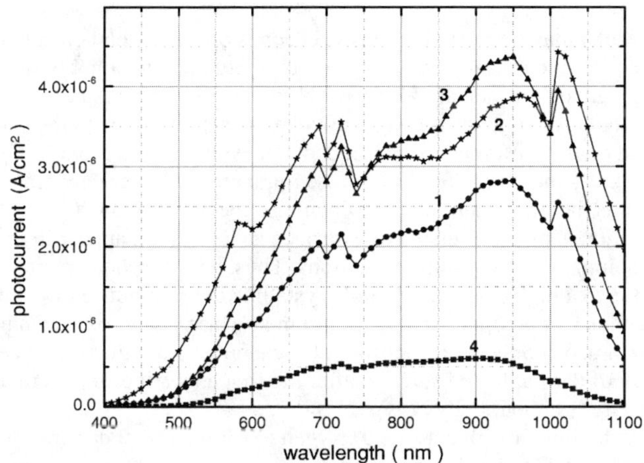

Figure 2. Spectral variations of the photocurrent of Cz samples implanted at 250 keV, annealed at 550°C for 1h before (1) and after etching in CP4 solution for 5s (2), 10s (3) and 12s (4).

annealing at 500°C. May be the ion accumulation zone is too close to the free surface, that enhances the out-diffusion of hydrogen (TRIM software indicates that R_p is about 0.25 µm and 0.35 µm below the free surface for 20 and 40 keV, respectively).

I-V and C-V curves reveal that an abrupt junction is formed after the first annealing at 350°C, which is progressively transformed in a graded one after subsequent annealing at higher temperatures. The doping level of the N-type zone is around 2.5×10^{17} cm^{-3}.

Scanning electron microscopy of a cross section of the implanted and annealed samples reveals the ion accumulation zone, which is located at the expected depth, i.e. R_p, of about 1 µm below the surface when the ion energy is 120 keV, as shown by Fig. 4. Etching the sample up to that thickness, we observe the formation of small etch pits, probably related with the formation of the well known blisters, which indicate that the region close to R_p is a defect- containing one.

Solar cells were realized with implanted wafers using aluminium grids and layers on the front and back surface, respectively. Although such cells are characterized by a high series resistance (1 Ω for an area of 4 cm2), certainly due to the perturbed zone around R_p and the free surface, under AM 1.5 sun light we have measured an open circuit photovoltage of 510 mV, a short circuit photocurrent of 21 mA/cm^2 and a saturation current of 3×10^{-9} A/cm^2. The preceding results are not of interest for photovoltaic devices, however they confirm the formation of a n-p junction in a P-type wafer.

DISCUSSION

The preceding results may be related to drastic changes of the region beneath the hydrogen-implanted surface in which the ion accumulation is so high that nanobubbles are formed, especially after an annealing step at 350°C. An N-type region appears beneath the surface which gives rise to an n-p junction and a photovoltaic behaviour, due to the generation of donors. This behaviour is enforced by subsequent annealing steps up to 550°C for samples implanted at 250 keV, and up to 500°C for those implanted at lower temperatures. The main requirement is the implantation dose which must be higher than 10^{16} cm^{-2}.

The donor generation is directly related to hydrogen as it does not appear in the samples implanted with helium under the same conditions. Thus, an amorphous region, or an imperfect one, due to a high concentration of crystalline defects induced by implantation (which could generate a potential barrier close to the surface), cannot explain the results.

A compensation of acceptors near the surface also cannot explain the results because the surface resistivity of the implanted surface, after the first annealing step, is relatively low and the conductivity is undoubtedly N-type.

The interaction of hydrogen with dissolved oxygen to form thermal donors could be assumed in oxygen rich Cz samples, however in FZ samples, with a low oxygen concentration, such an explanation is no longer valid, because similar effects are observed (there is only one difference: the photocurrents in Cz samples are just slightly increased by the annealing at temperatures above 350°C, while they are enhanced in FZ wafers).

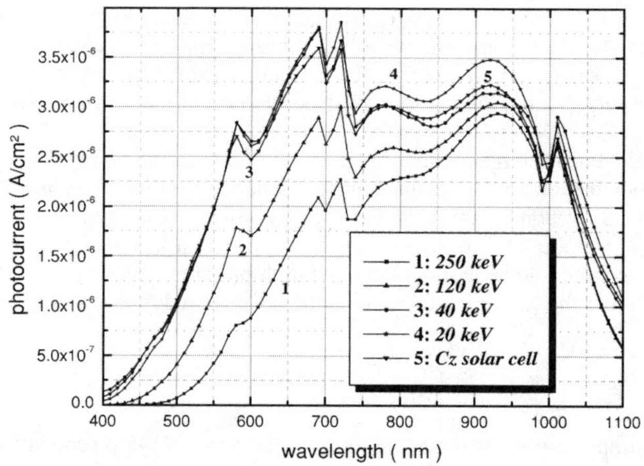

Figure 3. Spectral variations of the photocurrent of Cz implanted samples at various energies: (1) 250 keV, (2) 120 keV, (3) 40 keV and (4) 20 keV. Curve 5 is that of a classical Cz solar cell.

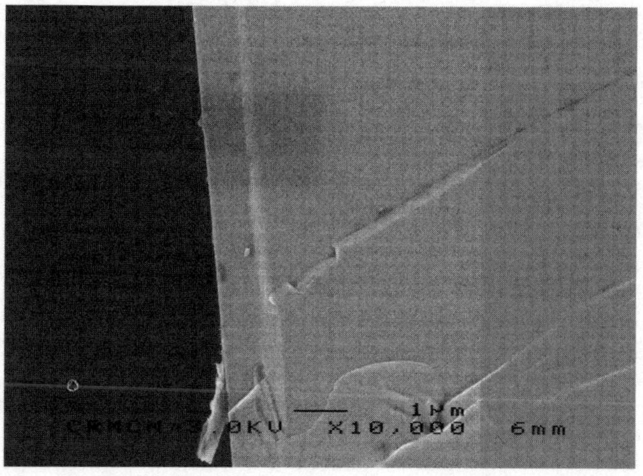

Figure 4. Scanning electron microscopy image of the cross section of a sample implanted at 120 keV and annealed at 500°C for 30 min.

Moreover, it was reported that the donors formed by hydrogen-oxygen interaction, at 450°C, are very stable [7]. Our results show that the N-type appears just after a first annealing step at 350°C and around 550 °C, the samples are converted back to their original P-type conductivity. Such temperatures correspond to the well known out-diffusion temperature range of hydrogen in silicon.

Thus it is believed that the most likely explanation of the preceding results is the formation of shallow donors beneath the implanted surface at a depth which corresponds to the projected range. Such donors are related to the high dose of hydrogen and may be due to the formation of nanobubbles in FZ, like in Cz, implanted silicon wafers. As annealing steps are required, it could be assumed that the high pressure which is achieved in the nanobubbles before the formation of nanocavities is involved in the donor formation mechanism.

CONCLUSION

Hydrogen ion implantations at doses higher than 10^{16} cm^{-2} in P-type crystalline silicon convert the near-surface region to N-type, due to the formation of shallow donors. After low temperature annealing steps and N-P junction is formed. The donor density increases with annealing temperatures in the range 350 to 500°C and reaches 2 to 3×10^{17} cm^{-3}. After annealing at temperatures above 600°C, the samples are converted back to their original P-type. The observed shallow donors are related to hydrogen.

ACKNOWLEDGEMENTS

The authors would like to thank Dr. Reedy (NREL-Boulder-Colorado-USA) for SIMS analysis, Professor Dallaporta and co-workers (Université de la Méditerranée-Marseille-France) for electronic microscopy imaging and Dr. Torregrosa (Ion Beam Services – Rousset-France) for hydrogen implantations.

This work was supported by Conseil Général 13-France, by Conseil Régional Provence, Alpes, Côte d'Azur – France, by CNRS(ECODEV) and by ADEME-France.

REFERENCES

1. S.J. Pearton, J.W.. Corbett and M. Stavola, in *Hydrogen in Crystalline Semiconductors*, edited by H.-J. Queisser, Springer Series in Materials Science 16, Berlin, 1992
2. K. Irmscher, H. Klose, K. Maas and J. Phys. C 17, 6317 (1984)
3. Y.V. Gorelkinskii and N.N. Nevinnyi, Nucl. Instr. Meth. 209/210, 677 (1983)
4. Y. Ohmura, Y. Zohta and M. Kanagawa, Phys. Stat. Sol. A 91, 509 (1985)
5. B.N. Mukashev, M.F. Tamendarov and S. Zhtakmoldin, Phys. Stat. Sol. A 91, 509 (1985)
6. Y.V. Gorelkinskii, V.O. Sigle and Z.S. Takivuev, Phys. Stat. Sol. A 15, K55 (1974)
7. R. Job, J.A. Weima, G. Grabosch, D. Borchert, W.R. Fahner, V. Raiko and A. G. Ulyashin, Solid State Phenomena, 69-70, 551 (1999)

Impact of Hydrogen Plasma Treatment on Gettering by He Implantation-Induced Cavities in Silicon

D. Alquier[1], E. Ntsoenzok[2], C.L. Liu[1,2], A. Vengurlekar[3] and S. Ashok[3]

[1] LMP, 16 rue Pierre et Marie Curie, BP 7155, F37071 Tours CEDEX, France
[2] CERI-CNRS, 3A rue de la Férollerie, 45071 Orléans CEDEX, France
[3] Department of Engineering Science and Mechanics, the Pennsylvania State University, 212 Earth and Engineering Science Building, University Park, PA 16802, USA

Abstract

The use of gettering techniques, with precise location of the gettering regions, has become crucial for device manufacturing. Helium-induced cavities have been shown to getter metallic impurities very effectively, but suffer from the drawback of requiring relatively high He doses. In this work, He-implanted Cz wafers of varying resistivity were subjected to hydrogen plasma hydrogenation prior to the cavity-formation anneal. We focus our studies on the cavity layer interactions with metal impurities. XTEM images reveal that hydrogenation increases the cavity radius. Our SIMS results show that the doping level has no influence on the metal gettering efficiency while the addition of plasma hydrogenation tends to decrease it. However, the efficiency can be controlled with the cavity radius which is interesting for future applications of the technique.

Introduction

The light elements, hydrogen and helium, are commonly used in semiconductor technology. Hydrogen exhibits extraordinary chemical activity in silicon, reacting with point and extended defects, surfaces and impurities. Effects such as passivation of pre-existing defects, and formation of new shallow and deep centers due to the presence of hydrogen in silicon are among the many well-known phenomena [1,2]. Helium is also used in silicon for gettering and minority carrier lifetime control applications [3-6].

High dose He implantation followed by thermal anneal results in the formation of a buried layer of nm-size cavities, whose large internal surfaces act as excellent gettering sites for fast-diffusing metallic impurities[3,4]. These defects strongly interact not only with metal impurities, but also with the dopant, affecting significantly their final profile and probably the electrical behavior of the device [4,5]. A clear understanding of the cavity formation and evolution is then required in order to safely introduce them in a device component. While many studies have shown that the presence of interstitial-type defects can lead to the shrinkage of He-cavities due to the interstitial capture by the cavities [6-8], there have so far been few reports dealing on the interaction of He-cavities with vacancies [9]. Plasma hydrogenation is widely encountered in semiconductor technology during various etching, deposition and passivation steps. It is well known that both defects and hydrogen atoms are injected into the subsurface layer under these conditions [10,11]. In previous work, we have demonstrated that additional H plasma treatment greatly impacts the morphology and size distribution of the He cavities induced by MeV implantation [9]. It is then crucial to figure out the impact of the hydrogen plasma step on the gettering efficiency of the cavities.

In this work, we present first the cross-sectional TEM results on the 160 keV He implantation induced cavities with and without an additional plasma hydrogenation. Then, we study the impact of the doping level on metal (copper) gettering. Finally, we discuss the impact of the hydrogenation process on the defect layer gettering efficiency.

Experimental

P type, <111>, CZ silicon wafers doped at $1x10^{15}$ and $1x10^{18}$ cm^{-3} were used for this study. Helium implantation was performed at room temperature at an energy of 160 keV and a dose of $5x10^{16}$ He^+/cm^2. The different dopant levels help evaluate the impact of the dopant concentration on the defect layer. The concentrations, however, were kept below the solid solubility in the studied temperature range to avoid precipitation effects [12]. Half of the samples were then subjected to an electron cyclotron resonance (ECR) high-density hydrogen plasma treatment. The plasma conditions are as follows: 600 W microwave power, 30 min. hydrogenation time and 250 $^{\circ}$C substrate temperature. After He implantation and with or without plasma hydrogenation, the samples were annealed. Furnace Annealing (FA) was performed at 800°C for 1h in a nitrogen gas ambient in order to form the cavity band. One set of the samples was contaminated on the backside by a spin-coated Copper source before the FA step. The evolution of He-cavities was studied by cross-sectional transmission electron microscopy (XTEM), using a Philips CM12 electron microscope. Secondary Ion Mass Spectrometry (SIMS) was used to detect the metal distributions before and after the annealing steps. All the measurements were made with an IMS-4F CAMECA system under appropriate conditions to avoid artifacts. These measurements enable us to evaluate the gettering efficiency for the metal.

Results and Discussion

The implantation of He at 160 keV for a dose of $5x10^{16}$ He^+/cm^2 followed by an annealing at 800°C for 1 hour creates a well-defined cavity band at a depth of 0.85 μm from the surface. The corresponding cavity bandwidth is 150 nm. A typical XTEM image of He-induced cavities is presented in Fig.1(a). From this image, it is clear that large cavities are distributed on both sides of the band. The cavity size distribution is in the range of 4.0 – 26.0 nm with an average diameter of about 11.0 nm. Moreover, other extended defects, interstitial type such as "rod-like" defects and dislocation loops, can also be seen beneath the cavity band.

As noted above, the addition of a hydrogen plasma treatment can largely influence the evolution of He-induced cavities. Indeed, in previous work on MeV He implantation-induced cavities [9], we have shown that plasma hydrogenation increases the mean size of the cavities and decreases their density. Similar results are obtained in the case of 160 keV He implanted Si samples, as shown in Fig. 1. The XTEM observation of the hydrogenated 160 keV He implanted sample annealed in the same condition is presented in Fig. 1(b). In this figure, one can observe that cavity size has increased. The enhancement in growth of cavities is clearly seen by size measurements. The cavities have diameters ranging from 4 – 30 nm with a mean value of about 14 nm. Meanwhile, large cavities with size up to 50 nm are also present. A reduction of the cavity bandwidth is also observed; from 150 without H-treatment to 100 nm with hydrogen. Two mechanisms are invoked to explain this phenomenon. First, the possible in-diffusion of H [11] can impact the cavity evolution as already demonstrated in the case of wafer splitting/bonding techniques [13,14]. Secondly, plasma hydrogenation also produces vacancy and vacancy-type defects in a region up to 1 µm deep from the surface [9]. One can easily surmise that part of these defects will interact with the He implanted defects. Both mechanisms can also coexist and we are carrying out experiments to discriminate between the two. Whatever be the exact mechanism, the additional hydrogen plasma treatment is effective in promoting the growth of the He-cavities.

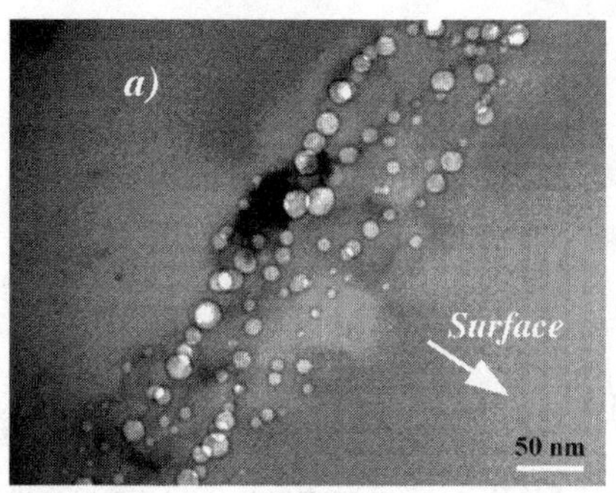

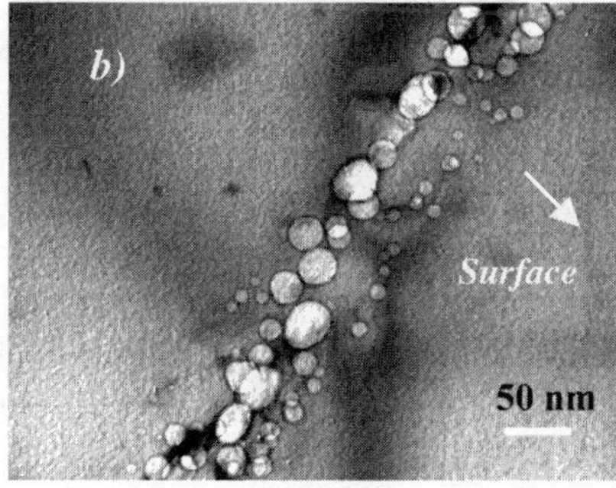

Figure 1: XTEM images of 160 keV He implanted Si samples at a dose of 5×10^{16} cm^{-2} and annealed at 800°C for 1 hour (a) without and (b) with additional hydrogen plasma treatment

As mentioned in the introduction, He implantation is an interesting alternative for gettering applications. In the case of power microelectronics, the wafer doping level can vary by three to four orders of magnitude. It is then crucial to know whether the doping level has an influence on metallic trapping or not, in order to safely incorporate this step in an industrial device application. Indeed, the metallic contaminants, even at low concentration, degrade device performance. A dopant level dependence of Cu segregation could have serious adverse effects on the devices. In figure 2, we present the Cu SIMS profiles obtained after an 800°C - 1h gettering treatment on wafers doped with 1×10^{15} B per cm^3 and 1×10^{18} B per cm^3. It is clear from figure 2 that the trapping of Cu occurs preferentially in the cavity band. We can also observe that the two samples have the same bandwidth located at an identical depth in the sample. To a first order, no modification of the Cu profile is observed when increasing boron concentration. Thus the gettering efficiency for Cu is unaffected by the boron doping level. Nevertheless, it is interesting to see in more detail the shape of the Cu profile in the cavity band. At the lower doping level, the Cu profile is almost Gaussian, corresponding to the cavity repartition usually observed in these conditions. When the doping level increases, we find that the repartition of Cu cannot be perfectly fitted by a symmetrical Gaussian distribution. The maximum is not, in this case, centered in the middle of the cavity band but seems to move deeper. This is consistent with the reported observation on 40 keV He induced-cavity evolution with the doping level [13]. According to this work, at increasing doping levels, the cavity repartition changed from a Gaussian curve to a curve with a maximum at the edge, just as in our Cu SIMS profile.

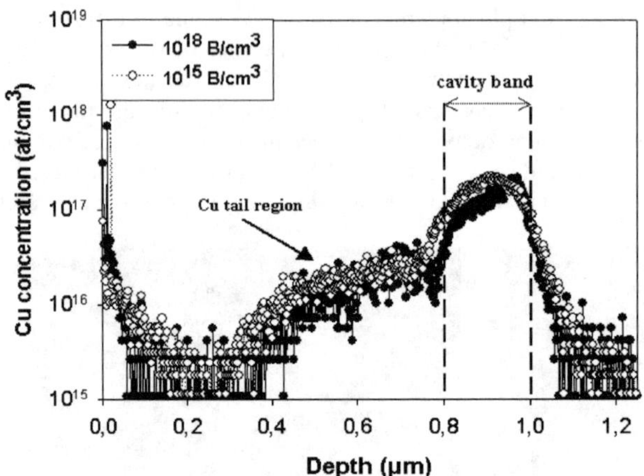

Figure 2: Comparison of cupper SIMS profiles after the 160 keV He gettering stage at 800°C for 1h on B-doped wafers at $1x10^{15}$cm^{-3} (o) and 10^{18}cm^{-3} (•)

Moreover, a tail region is also observed in Cu SIMS profile in figure 2. This region is always observed with Cu contamination (see for example [2,14]). Some authors have attributed the tail region to Rp/2 defects [14]. However these defects are not seen in TEM pictures [see figure 1(a)] and this tail region is also not observed in the case of other contaminants such as iron [5,15]. Finally, one can figure out that the doping level does not impact gettering efficiency which is highly important for device applications. The same observation was done on samples with additional H plasma.

In order to investigate the influence of the plasma hydrogenation on the copper gettering into the He-induced defect layer, another set of experiments was performed. Low-doped wafers were implanted with helium ($5x10^{16}$ He per cm^2 at 160 keV). Half of the wafers were then subjected to H plasma as previously described. A single plasma hydrogenation step was also applied on others samples. All the wafers were then furnace annealed after a Cu backside contamination step. In figure 3, we show the Cu SIMS profiles obtained after an 800°C-1h anneal on a $1x10^{15}$ cm^{-3} B-doped wafer, with and without hydrogenation. Significant differences are evident between the two samples.

First, a Cu peak is seen close to the surface region (zone 1). A similar peak, located at the same position but with a slightly higher amplitude, is observed in a sample that has been only hydrogenated (no He implantation). This peak is probably due to extended defects resulting from the H plasma and anneal. Due to the hydrogenation, we assume these defects to be vacancy type. Moreover, Stritzker et al have show that Cu getters preferentially or even exclusively in vacancy type defects when both types of extended defects are present [15].

In zone 2, one can clearly observed the impact of the plasma step. The Cu SIMS profiles exhibit similar shapes (Gaussian curve + tail region) with differences in thickness and amount of gettered Cu. Indeed, figure 3 shows that the hydrogenated plasma sample presents a narrower band of gettered copper, corresponding to the cavity bandwidth decrease observed by TEM in

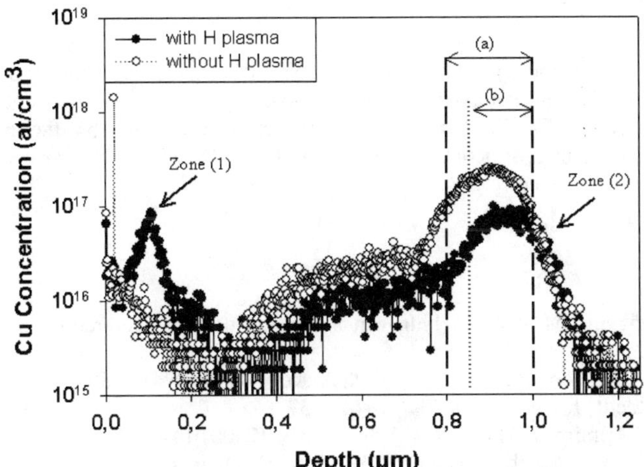

Figure 3: Comparison of cupper SIMS profiles after the gettering stage at 800°C for 1h with (●) or without (○) additional plasma hydrogenation.

figure 1. Concerning the gettering efficiency of the cavity band, one can observe that the amplitude of the Gaussian distribution has decreased by a factor of 4. This can be related to the TEM observations where we have found that the average radius has increased largely with plasma hydrogenation. A study on the influence of cavity curvature on gettering shows that small cavities are much more efficient than larger ones [16]. They have shown that the efficiency decreases by a factor of 4 when the average radius rises from 7nm to 15nm, which is consistent with this study, if we consider that the cavity density decrease also has to be taken into account in our case. Nevertheless, the addition of a plasma hydrogenation that ensures a cavity size increase as expected, impacts the metal gettering efficiency in a controlled way that can be monitored by the cavity radius. This technique can be useful for He dose reduction while conserving gettering efficiency. Additional efforts are needed to avert surface defect area (zone 1).

Conclusions

The creation of a nanocavity band by helium implantation is an interesting option for gettering of metal impurities. Additional plasma hydrogenation, through the injection both defects and hydrogen, can influence the cavity evolution. In this work, we proceed to various experiments to investigate the interaction of cavity band and metal with and without the additional H-plasma treatment. We have shown that doping level does not influence the gettering efficiency. Our TEM measurements show that plasma hydrogenation tends to increase the cavity size. However, SIMS profiles of the gettered Cu show decreased gettering efficiency with this step. This decrease can be controlled or even suppressed by changing processing parameters such as dose, anneal temperature and plasma conditions. In this way, plasma hydrogenation may still be an effective technique to reduce the He implant dose, as desired for a viable gettering step in industrial applications.

Acknowledgments

The authors want to thank Ch. Dubois (LPM at INSA Lyon) for her generous support on SIMS measurements. The authors from the Pennsylvania State University wish to acknowledge use of the Penn State Nanofabrication Facility, and partial support of the work by the National Renewable Energy Laboratory under subcontract number ACQ-9-29639-04. This work was also supported by Le Studium (Region Centre).

References

1. S.J. Pearton, J.W.Corbert and M.Starola, Hydrogen in crystalline semiconductors (Springer-Verlag, Heidelberg, 1992)
2. E.O.Sveinjornsson, G.I.Anderson and O.Engstrom, Phys. Rev. B49 (1994) p7801.
3. S.M. Myers, D.M. Follstaedt, J. Appl. Phys. 79 (1996) 1337.
4. V. Raineri, M. Saggio, E. Rimini, J. Mater. Res.,Vol. 15, N°7 (2000) 1449.
5. F. Cayrel, D. Alquier, L. Ventura, L. Vincent, F. Roqueta, C. Dubois, R. Jerisian, GADEST 2003, Zeuthen, Germany, Sept. 21-26 2003 (invited paper) , Solid State Phen. Vol. 95-96 (2004) p 297
6. S. E. Donnelly, V. M. Vishnyakov, R. C. Bitcher and G. Carter, Nucl. Instru. Meths. B175-177 (2001) p132.
7. V. Raineri and S. U. Campisano, Appl ; Phys. Lett. Vol. 69 (1996) p1783.
8. Changlong Liu, E.Ntsoenzok, R. Delamare, D. Alquier, and G. Regula, L. Vincent, C. Filadelfo and A. Claverie, Mat. Sci. & Eng. B, (2003), in press.
9. C.L.Liu, E. Ntsoenzok, M. F. Barthe, P. Desgardin, S. Ashok, A. Vengurlekar, D. Alquier, M. -O. Ruault, Solid State Phenomena, Vol. 95-96, 307-312 (2003).
10. H. Schut, A. Van Veen, S. W. H. Eijt, R. Job, A. G. Ulyashin, W. R. Fahrner, Nucl. Instru. Meths. B 186 (2002) p94.
11. A. G. Ulyashin, R. Job, W. R. Fahrner, D. Grambole, F. Herrmann, Diffusion and Defects Data Pt. B: Solid State Phenomena, 82-84, (2002) p315.
12. D. Alquier, F. Roqueta, L. Ventura, F. Cayrel, C. Dubois and R. Jérisian, Jap. J. Appl. Phys. Vol (41) pp 3625-3628 Part 1 N° 6A June 2002
13. F. Cayrel, L. Vincent, D. Alquier, F. Cristiano, L. Ventura, C. Dubois and A. Claverie, GADEST 2003, Zeuthen, Germany, Sept. 21-26 2003, Solid State Phen. Vol. 95-96 (2004) p 325
14. A. Peeva, P.F.P Fitchner, M. Behar, R. Kogler, W. Skorupa, Nucl. Instru. Meths. B 175-177 (2001) p 176
15.B. Stritzker, M. Petravic, J. Wong-Leung and J.S. Williams, Appl. Phys. Lett. Vol. 78 (18) (2001) p.2682
16. F. Schiettekatte, C. Wintgens and S. Roorda, Appl. Phys. Lett. Vol. 74 (13) (1999) p.1857

How much hydrogen and voids are energetically stable in silicon thin films?

Anna Fontcuberta i Morral, Holger Vach, Pere Roca i Cabarrocas
LPICM (UMR 7647), Ecole Polytechnique, 91128 Palaiseau, Cedex France

Abstract

We have developed a model to account for the effects of hydrogen and voids on the structural stability of silicon thin films. The model is based on both experiments and theory. First, hydrogenated amorphous silicon films (a-Si:H) with various hydrogen contents were obtained by Plasma Enhanced Chemical Vapor Deposition. A linear correlation between hydrogen content and void fraction was observed. By tuning the deposition conditions, polymorphous silicon films with hydrogen contents up to 15%, very small void fractions (0.5%) and excellent electronic properties were also obtained. Density Functional Theory (DFT) calculations were performed to determine the formation energy for four types of silicon tetrahedra of the form $Si-Si_nH_{4-n}$ (n=1, 2, 3, 4). In our model, these tetrahedral units are considered as the building blocks of the silicon thin films. Considering a homogeneous distribution of hydrogen in the solid, the proportion of the different $Si-Si_nH_{4-n}$ tetrahedra as a function of the hydrogen concentration was calculated. Then, the formation energy of hydrogenated amorphous silicon (a-Si:H) was calculated as a function of the hydrogen content and for various porosities. The model predicts that hydrogen incorporation does render the a-Si:H structure unstable for different hydrogen contents depending on the void fraction. Our results show that polymorphous silicon films with hydrogen concentrations up to 15% can be as stable as standard amorphous silicon with 2% hydrogen content, provided that the presence of hydrogen is not associated with the incorporation of porosity in the film.

INTRODUCTION

Amorphous silicon was deposited for the first time in 1879 by Ogier[1]. 90 years later, this material was re-discovered by Chittick et al. who showed that in spite its disordered structure, amorphous silicon did behave like a semiconductor[2,3]. This material became especially important after the acceptation of the fundamental role of hydrogen in the passivation of dangling bonds[4]. Since then, a lot of work has been devoted to the understanding of the different roles of hydrogen in silicon. In particular the Staebler-Wronksi effect, which consists in the increase of the density of dangling bonds upon illumination of a-Si:H[5] and in which it is believed that hydrogen plays a significant role. It is also generally accepted that a large hydrogen content in a-Si:H leads to poor electronic properties. Large efforts have been devoted to the obtaining of a dense a-Si:H with very low hydrogen content[6]. However, we have recently shown that polymorphous silicon thin films with high hydrogen content and good electronic properties can be obtained[7, 8], which shows that large hydrogen contents are not incompatible with good electronic properties of silicon thin films.

The solubility of hydrogen in silicon has been calculated in the past by applying thermodynamical models[9, 10]. In these models a-Si:H was considered as an alloy of hydrogen and silicon. They predicted a solubility between 4 and 8% of hydrogen in silicon, meaning that silicon thin films with higher hydrogen content are structurally unstable and tend to undergo phase transformations which result in a diffusion of hydrogen out of the material. Experimentally, it is generally observed that silicon films with such low hydrogen contents usually present excellent electronic properties. However, these models consider a-Si:H as a mixture of Si and H, consisting of just Si-Si and Si-H bonds and ignoring the tetrahedral geometry of the bonds around each silicon atom. Moreover these models do not include any correlation of hydrogen content with the void fraction of the film, which exists and is related to the deposition mechanisms of a-Si:H. In this work we present an analysis of the structural stability of hydrogen in silicon, based on experimental measurements of hydrogen content and mass density, as well as on theoretical calculations which take into account the tetrahedron geometry of the bonding around the silicon atoms.

EXPERIMENTAL AND THEORETICAL CALCULATIONS

Silicon thin films were deposited by Plasma Enhanced Chemical Vapor Deposition. Standard hydrogenated amorphous silicon (a-Si:H) was obtained by the decomposition of pure silane at low pressure and low RF power (6 Pa and 17 mW/cm^2). In order to vary the hydrogen and void contents, a series of films was deposited at different substrate temperatures ranging from 50°C to 300°C. Polymorphous silicon[11] films were produced by the dissociation of 3% silane in hydrogen gas mixtures under an RF power of 100 mW/cm^2. The total gas pressure was varied between 120 and 240 Pa. Rutherford Backscattering Spectroscopy (RBS) and Elastic Recoil Detection Analysis (ERDA) measurements were performed to obtain the density and hydrogen content of the films. For an accurate determination of the density, the films were deposited on a 400 Å thin chromium layer thermally evaporated on a crystalline silicon substrate.

The tetrahedron model[12, 13] was applied for the in both the theoretical mass and energy density calculations. In this model, the network of a-Si:H is assumed to consist of an ensemble of voids and silicon centered tetrahedra, whose vertexes can be either silicon or hydrogen atoms, and symbolized by: Si-Si$_n$H$_{4-n}$ (n=1, 2, 3, 4), SiH$_4$ being excluded from the model. Density functional theory was applied for the calculation of the energy formation of the different building blocks of a-Si:H.

RESULTS

Figure 1a shows the hydrogen content of the silicon thin films as a function of the void fraction, deduced from mass density obtained by RBS measurements[13]. In the case of standard a-Si:H, there is a linear correlation between the hydrogen content and the void fraction. However, the polymorphous silicon thin film has a hydrogen content of 16%, without any reduction of the mass density. Such high hydrogen content together with such low void fraction (0.5%) may seem contradictory from the point of view of standard amorphous silicon, but we'll show in the following that this is not the case. In Figure 1b the theoretical mass density is plotted as a function of the hydrogen content. The calculations were done using the tetrahedron model of amorphous silicon. Two cases are represented: The first one only takes into account the reduction of mass density due to the incorporation of hydrogen, while the second one considers the correlation found between the hydrogen content and the void fraction (Fig. 1a). As shown in Fig 1b, it is possible to incorporate up to 20 % hydrogen in the silicon network without any substantial reduction of the mass density.

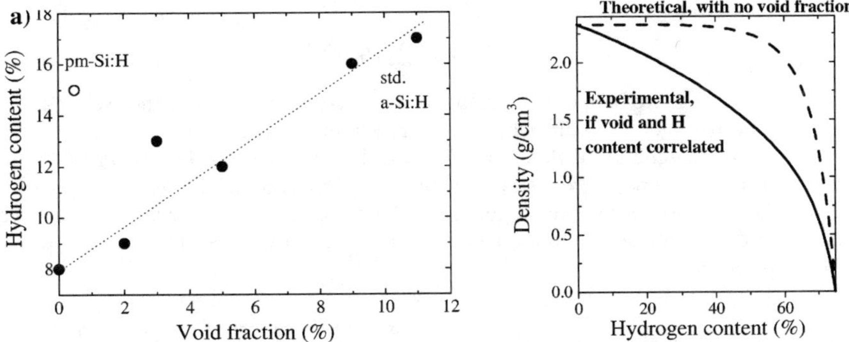

Figure 1 a) Experimental correlation between hydrogen content and void fraction of standard amorphous silicon deposited by RF-PECVD **b)** Comparison between the density of theoretical amorphous silicon without any correlation between hydrogen content and void fraction and experimental amorphous silicon.

For the calculations of the formation energy per unit volume of a-Si:H as a function of hydrogen content, basic building blocks were slightly different from the $Si-Si_nH_{4-n}$ (n=1, 2, 3, 4) used typically in the model. This is due to the fact that the basic units of the tetrahedron model have dangling bonds, which would lead into a wrong evaluation of the formation energy. The basic units were the molecules SiH_4, SiH_3-SiH_3, $SiH_2-(SiH_3)_2$ and $SiH-(SiH_3)_3$. The formation energy of each unit was calculated with the *Gaussian*[14] program suite using Becke's UB3LYP density functional. The results on the energy formation are shown in Figure 2a.

a)

Structure	Energy (eV)
SiH_4	-7941.83
SiH_3-SiH_3	-15851.77
$SiH_2-(SiH_3)_2$	-23761.74
$SiH-(SiH_3)_3$	-31671.73
$Si-(SiH_3)_4$	-39581.75

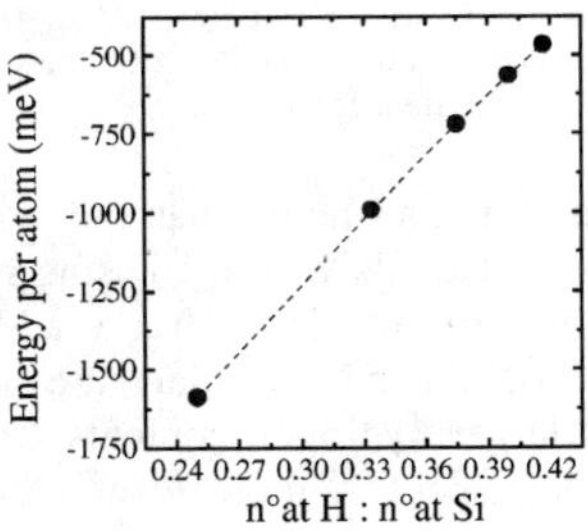

Figure 2 a) Values of the energy formation of the different tetrahedral structures calculated by DFT. **b)** From the table in Fig. 2a, energy formation per atom as a function of the ratio H/Si

In Figure 2b,calculated from the table in Fig. 2a, the formation energy per atom is plotted for the different hydrogen/silicon ratios of the molecules. From this plot, it is clear that the molecules with a higher hydrogen/silicon ratio are less stable. The values of formation energy of the basic units together with the tetrahedron model were applied to calculate the formation energy per unit volume of a-Si:H as a function of hydrogen content using the following expression:

$$\rho_E(H) = \rho \cdot {}^{N_A}\!\!/_{m_{Si}} \sum_n p_n(H) \cdot E_n$$

where N_A is the Avogadro Number, m_{si} is the atomic mass of silicon, p_n is the probability of presence of a n-hydrogenated unit (which is a function of the hydrogen content in the film) and E_n the formation energy of this unit. The calculation of energy density was done for two cases: In the first one, we consider no additional reduction of the mass density of a-Si:H with the incorporation of hydrogen and in the second case we consider the linear correlation of the hydrogen and void fraction found in standard a-Si:H (Fig.1a) . The results are shown in Fig. 3a:

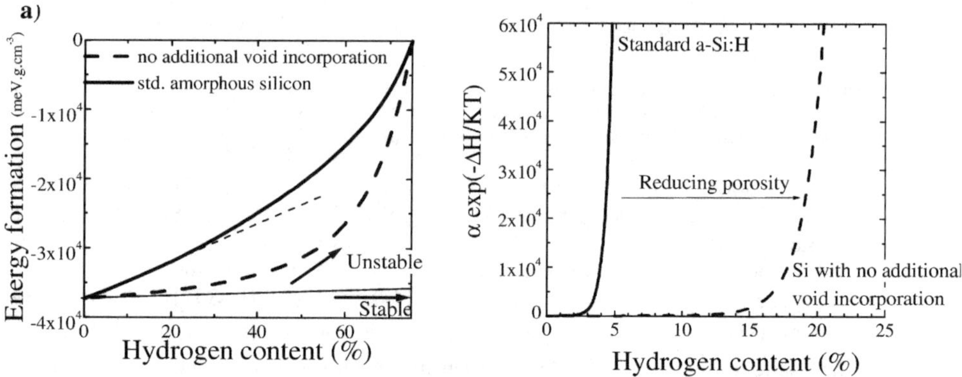

Figure 3. a) Formation energy density as a function of the hydrogen content for two kinds of amorphous silicon: one with the maximal theoretical density (no additional void incorporation) and another with the experimental density, which is correlated to the hydrogen content. **b)** Exponential of the formation energy, proportional to the probability of a phase transformation of hydrogenated silicon as a function of the hydrogen content, deduced from a).

The plot in Fig. 3a shows that the absolute value of the formation energy of a-Si:H decreases for materials with increasing hydrogen content. Comparing the case of standard a-Si:H to that of a-Si:H with no additional void incorporation (maximal theoretical density), we obtain that standard a-Si:H is the most unstable. From this figure one can deduce that larger hydrogen contents are more stable in dense a-Si:H than in standard a-Si:H. This statement is more clear when calculating the probability of the material to undergo a phase transformation to a more stable phase that would lead to the outdiffusion of hydrogen of the material:

$$r_t \propto \exp(-{}^{\Delta H}\!\!/_{KT})$$

where K is the Boltzman constant, T the temperature (300K for this calculation) and ΔH is the difference of energy between a final state of non-hydrogenated pure silicon and a initial hydrogenated state. From Fig 3b it is clear that there is an onset of hydrogen concentration above which the material has to undergo a phase transformation and let the hydrogen diffuse out of the material. The onset is about three times lower for standard amorphous silicon thin films with respect to dense silicon thin films. This result is in agreement with the fact that the good electronic properties of a-Si:H thin films are correlated with low hydrogen contents, but also shows that stable silicon films with hydrogen contents of 15% are possible.

CONCLUSION

As a conclusion, we have shown that the hydrogen content of standard a-Si:H is linearly correlated to the void fraction of the material. We have also shown that by tuning the deposition conditions it is possible to obtain dense films with hydrogen content up to 15%. Combining the tetrahedron model and density functional calculations, we have calculated the structural stability of a-Si:H as a function of the hydrogen content and void fractions. We have found that hydrogen can be incorporated up to about ~15% in amorphous silicon without destabilizing the structure. This results contradicts the general belief that such large hydrogen contents render a-Si:H unstable, but at the same time it is in good correlation with our experimental results on polymorphous silicon.

References

[1] J. Ogier, Bl. Soc. Chim. **8**, 116 (1879)

[2] R.C. Chittick, J.H. Alexander, H.F.Sterling, J. Electrochemical Soc. **116**, 77 (1969)

[3] P.G. LeComber, W.E. Spear, Phys. Rev. Lett. **25**, 509 (1970)

[4] D. Kaplan, N. Sol. Et G. Velasco, Appl. Phys. Lett. **33**, 440 (1978)

[5] D.L. Staebler, C.R. Wronksi, Appl. Phys. Lett. **31**, 456 (1977)

[6] A.H. Mahan, J. Yang, S. Guha, D.L. Williamson, Phys. Rev. B **61**, 1677 (2000)

[7] S. Vignoli, R. Butté, R. Meaudre, M. Meaudre, and P. Roca i Cabarrocas, J. Phys. Condens. Matter **11**, 8749 (1999).

[8] J.P. Kleider, C. Longeaud, M. Gauthier, M. Meaudre, R. Meaudre, R. Butté, S. Vignoli, and P. Roca i Cabarrocas, Appl. Phys. Lett. **75**, 3351 (1999)

[9] J. Robertson, J. Non-Cryst. Solids **266-269**, 79 (2000)

[10] S, Acco, D.L. Williamson, P.A. Stolk, F.W. Saris, M.J. van der Boogaard, W.C. Sinke, W.F. van der Weg, S. Roorda, P.C. Walm, Phys. Rev. B **53** 4415 (1996)

[11] P. Roca i Cabarrocas, A. Fontcuberta i Morral and Y. Poissant, Thin Solid Films Vol. **403-404**, 39 (2002)

[12] K. Mui, D.K. Basa, F.W. Smith, Phys. Rev. B **35** 8089 (1987).

[13] A. Fontcuberta i Morral, P. Roca i Cabarrocas, C. Clerc, Phys. Rev. B (2004)

[14] M.Z. Frisch et al., Gaussian 98, Gaussian, Inc., Pittsburgh, PA (1998)

METASTABLE DEFECTS IN a-Si:H and a-Ge:H: THE ROLE OF HYDROGEN

T. Su, P. C. Taylor, and J. Whitaker
Department of Physics, University of Utah, Salt Lake City, UT 84112-0830

ABSTRACT

Recent advances in understanding the role of hydrogen in the production and annealing of defects that contribute to the Staebler-Wronski effect in hydrogenated amorphous silicon (a-Si:H) are discussed. We discuss (1) the observation by nuclear magnetic resonance of a paired hydrogen site that stabilizes the silicon dangling bond defects that produce the Staebler-Wronski effect, (2) the observation of the Staebler-Wronski effect in a-Ge:H, and (3) the production of neutral silicon dangling bonds in tritiated a-Si:H.

INTDRODUCTION

The interaction between hydrogen and defects in crystalline and amorphous solids is of general interest in many diverse materials systems [1,2,3]. In particular, the roles of hydrogen both as a diffusing species that mediates the metastable defects and as a metastable defect itself remain controversial. Therefore, understanding the role of hydrogen in the production of metastable defects in hydrogenated amorphous silicon (a-Si:H) after irradiation with light [4], is essential in providing a general understanding of the role of hydrogen in mediating metastabilities in a wide class of materials. Specifically, it has been known for about 30 years that irradiation of a-Si:H with light of band-gap energy produces a decrease in both the photo- and dark conductivities [4]. This effect is known as the Staebler-Wronski effect after its original discoverers. From electron spin resonance (ESR) measurements [5], the defects responsible for the decreases in conductivity are known to be silicon dangling bonds. Hydrogen has long been invoked as important in stabilizing these dangling-bond defects, but the experimental proof of this conjecture has been elusive. We have reported an ^{1}H nuclear magnetic resonance (NMR) signal in a-Si:H that occurs only after light soaking for 600 hours [6]. This signal, which is attributed to a pair of hydrogen atoms, exhibits similar annealing kinetics to that of the defects created during light-soaking, and the concentration of these sites is comparable to that of the defects measured by electron spin resonance (ESR). The inescapable conclusion is that these paired hydrogen sites, which could either be formed or revealed (i.e., SiH_2 sites) on light soaking, stabilize the silicon dangling bond defects that cause the Staebler-Wronski effect. These experiments are described in the next section.

Although the Staebler-Wronski effect has been studied for many years in a-Si:H and in a-Si_xGe_{1-x}:H alloys with x > 0.8, it has only recently been observed in a-Ge:H [7]. Although this important defect has been studied in great detail in a-Si:H and a-Si_xGe_{1-x}:H alloys with x ≥ 0.8, the results in a-Ge:H have been mixed and controversial [8,9,10,11]. In a-Ge:H some studies have reported a decrease in the photoconductivity on irradiation with light [8,9,10] while others have reported an increase [11]. Even those that reported a decrease noted that the effect annealed at room temperature in the dark after several days. Several authors have even suggested that the Staebler-Wronski effect may not occur in a-Ge:H [12,13,14,15]. The suggestion is based on the idea that below a specific optical band gap energy there is insufficient recombination energy to lead to the creation of metastable defects. In the present work we show that this idea is incorrect. More importantly, we show that the discovery of a second elemental model system for studying the Staebler-Wronski effect, namely a-Ge:H, will eventually place serious restrictions on the microscopic models proposed to explain this important effect. These results are reviewed in a subsequent section.

Another well-known fact concerning the Staebler-Wronski effect is that the production of dangling bonds by light is very inefficient. The usual assumption is that very few recombination events produce defects, but it is also possible that non-radiative recombination creates defects very efficiently but that these precursor defects are stabilized only very inefficiently. The latter interpretation is supported by several experiments that suggest many more defects are created by light than the measured densities of dangling bonds [16,17]. To examine this question recent experiments [18,19] have investigated the intrinsic growth of defects in tritiated, hydrogenated amorphous silicon (a-Si:H,T) with approximately 6 at. % tritium. In this system, which provides an alternative approach to inducing defects optically, the tritium decays to He^3, emitting a beta particle, of average energy 5.7 keV, and an antineutrino. The half-life for this reaction is 12.5 years. In the tritiated samples each beta decay creates a defect, which may or may not be metastable, by converting a bonded tritium to an interstitial helium and thereby leaving behind a silicon dangling bond. The great advantage of this system is that we know exactly how many initial defects are created, and therefore one can track these defects through ESR and photothermal deflection spectroscopy (PDS).   The defects created by tritium decay anneal at approximately 150 C in a manner analogous to those created optically [18,19], but the defect densities are several orders of magnitude smaller than one would expect if each tritium decay produced a metastable silicon dangling bond. After one year, the decay of tritium to helium would be expected to yield defect densities of $\sim 10^{20} cm^{-3}$ without saturation, but the increases observed by ESR or PDS are much smaller [19]. These results are also reviewed in a subsequent section.

HYDROGEN AND THE STAEBLER-WRONSKI EFFECT

We have reported the first direct evidence for optically induced changes in the local environment of a subset of the hydrogen atoms *at densities comparable* to those of the silicon dangling bonds that contribute to the Staebler-Wronski effect [6]. We report here similar effects in samples of a-Si:H made by a technique (hydrogen dilution of silane in a PECVD reactor) that reduces the saturated densities of these defects. After optical excitation, the nuclear magnetic resonance (NMR) of ^1H exhibits a hydrogen "doublet" that corresponds to a site with two hydrogen atoms spaced approximately 2.3 Å apart. This hydrogen "defect", whose density is between 10^{17} and 10^{18} cm^{-3}, can be thermally annealed with kinetics that generally match the annealing of the silicon dangling bonds. Although a detailed microscopic picture of this defect remains elusive, the direct tie to hydrogen is a critical first step that places severe constraints on existing microscopic models for the Staebler-Wronski effect.

The films of a-Si:H were made in a large area deposition system by DC plasma enhanced chemical vapor deposition of silane (SiH_4) [6]. The pressure in the reactor was 0.5 Torr; the power density was ~ 50 mW/cm^2; and the substrate temperature was ~ 200 C. These conditions represent those used in standard photovoltaic devices for which the electronic and optical properties are very well known. The films were deposited on Al foil and half of the area of each film was irradiated with a solar simulator for 600 hr (light-soaked sample). Four NMR samples were made from the irradiated and un-irradiated areas by dissolving away the Al foils in dilute hydrochloric acid and placing the cleaned powders in quartz tubes. The NMR measurements were made near 7 K using a standard pulsed spectrometer as described elsewhere [20]. The stimulated echo from a Jeener-Broekaert three-pulse sequence [21] was employed to measure the ^{1}H NMR signals. This pulse sequence probes the decay of dipolar order among the hydrogen nuclei in the sample. Since the amplitude of the stimulated echo depends sensitively on the local environments of the hydrogen atoms or molecules, one can emphasize specific hydrogen sites by choosing the two pulse separations judiciously. This technique has been used effectively to measure the concentrations of hydrogen molecules in films of a-Si:H [20]. At large values of the separation between the second and third pulses one can emphasize any hydrogen site that interacts only weakly with silicon dangling bonds. After measuring the light-soaked state, the samples were annealed in a nitrogen environment.

In Fig. 1 we show the standard ^{1}H NMR lineshape for the hydrogen bonded to silicon in a-Si:H. The Zeeman frequency has been subtracted from this spectrum so that zero represents the NMR response in the absence of any solid-state interactions. The slight positive/negative asymmetry in the figure is an artifact due to phase errors in the Fourier transform of the data. The narrow Lorentzian line, whose full width at half maximum is approximately 3 kHz, is attributed to hydrogen atoms bonded to silicon atoms at random positions in the amorphous lattice. The broad line, whose full width at half maximum is approximately 25 kHz, is attributed to hydrogen atoms bonded to silicon atoms in a clustered environment, such as would occur at a "divacancy" or small void in the lattice. As mentioned in the previous section, the Jeener-Broekaert pulse sequence greatly suppresses these lines.

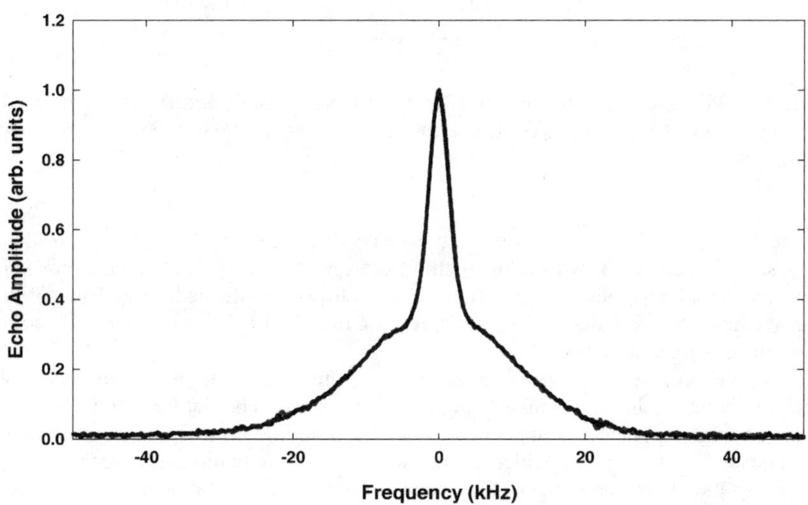

Figure 1. ^{1}H NMR lineshape for hydrogen bonded to silicon in a-Si:H. See text for details.

In Fig. 2 we show the Fourier transforms of the stimulated dipolar echoes, as obtained from the Jeener-Broekaert pulse sequence, in the light soaked sample of a-Si:H made without hydrogen dilution. From top to bottom, the traces are obtained at 20, 10, 7, and 6.5 K, respectively. The features at about ± 8 kHz are due to the paired hydrogen sites that appear only in the light soaked sample. In the half of the film that was not exposed to light, these features do not occur.

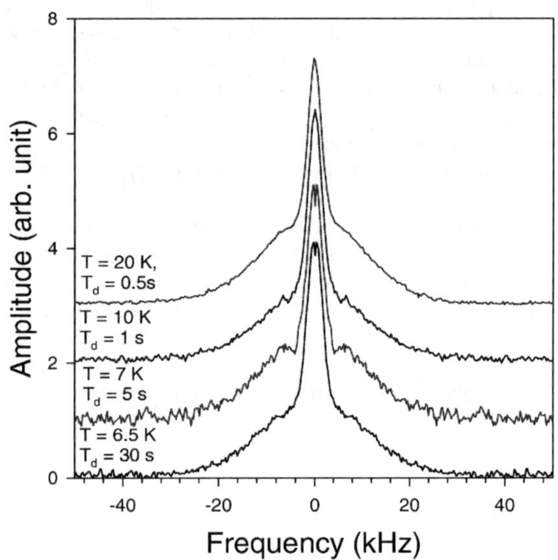

Figure 2. NMR lineshapes for ^{1}H from the light-soaked sample at different temperatures. The magnetic field is 3.5 T. The two dashed lines indicate where the hydrogen doublet appears. See text for details.

After annealing at 200 C for 4 hours the two peaks disappear, and the spectrum is identical to that observed in the as-grown sample within experimental error. This procedure is known to anneal the Staebler-Wronski degradation observed in solar cells made from films of this type [22]. We have thus found a distinct hydrogen site in a-Si:H that appears on light soaking and disappears on annealing.

Further conclusions require progressively more interpretation. Figure 3 is a subtraction of the lineshapes in the light-soaked sample at 7 and 10 K. This subtraction yields an estimate of the additional component that appears only on light soaking. This figure shows that the lineshape is indeed due to a doublet and not something more complicated such as a triplet where intensity would also appear at zero frequency. Once we have established that the site is a doublet, the spacing between the paired hydrogen atoms, R, can be estimated. Because the splitting depends inversely on R^3, the separation can be estimated fairly accurately as $R_o = 2.3 \pm 0.2$ Å.

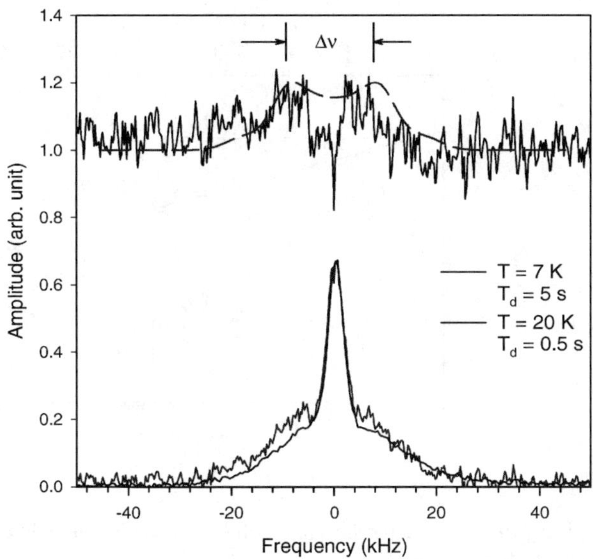

Figure 3. Bottom: NMR echo lineshapes at 7 K (noisy curve) and 20 K in a-Si:H. Top: Subtraction of the 7 K curve from the 20 K curve to estimate the hydrogen doublet lineshape. The dashed line is a fit to the data as discussed in the text.

One can also estimate the density of these optically induced hydrogen doublets. Figure 4 shows the decay of the echo intensity as a function of τ_2 for the two bonded hydrogen lines and the metastable doublet. The relative densities of these three contributions to the NMR can be estimated by extrapolating the respective echo amplitudes to zero separation between the three pulses (as τ_1 and τ_2 both approach zero; only the τ_2 dependence is shown in Fig. 4). These extrapolations work because the echo intensity extrapolated to zero time is transformed by the Fourier transformation into the integrated intensity in frequency space. There is considerable error in extrapolating the doublet signal as τ_1 and τ_2 approach zero, and therefore, the density of the paired hydrogen sites can only be estimated to be between 10^{17} and 10^{18} cm^{-3}. Although the error large, this range encompasses the density of optically induced silicon dangling bonds as measured by ESR. The ESR spin density is 5×10^{16} cm^{-3}, but this number is also uncertain by about a factor of five. The ESR signal is reduced by more than a factor of five on annealing at 200 C. This reduction is consistent with the complete loss of the doublet signal on annealing.

If the densities of silicon dangling bonds and hydrogen atoms contributing to the doublet are in fact the same, then this equality answers the long standing question as to why the silicon dangling bonds that are stable (dark ESR signal) and those that are metastable (optically induced ESR signal) have identical lineshapes. The answer is that they are, indeed, the same, but one can only anneal up to density that is twice the density of metastable hydrogen doublets. In addition, it is well known that after many inducing and annealing cycles the Staebler-Wronski effect is not completely reversible [23,24], but instead the stable dangling bond density increases gradually.

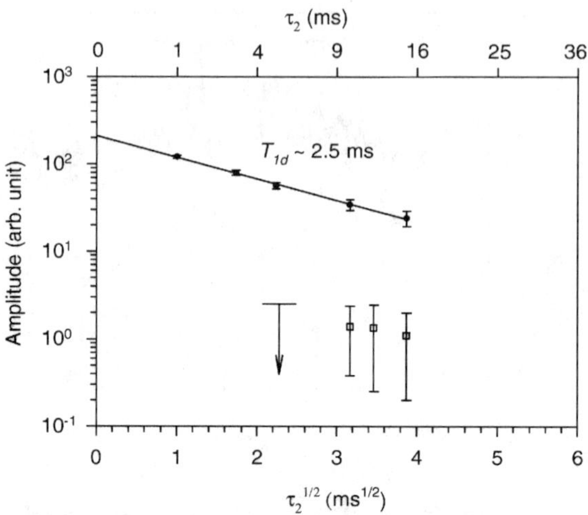

Figure 4. Dipolar decay of the bonded hydrogen (solid circles) and the paired hydrogen site (open squares). See text for details.

This fact can also be explained since one can easily imagine that on rare occasions the hydrogen atoms might come together in such an environment as to form a stable hydrogen molecule trapped in a microvoid.

We have performed similar experiments on samples of a-Si:H that were made using hydrogen gas to dilute the silane gas during growth. This procedure is known to decrease the degradation considerably in solar cells. The doublet also appears in these samples after light soaking as shown in Fig. 5. The top trace represents the half of the film that was irradiated in the solar simulator, and the bottom trace represents the unirradiated half of the film. The two arrows indicate the positions of the doublet peaks in the undiluted sample shown in Figs. 1 through 3. As expected the signal-to-noise ratio of the doublet is not as good because the density of doublet sites is smaller by at least a factor of three.

The interaction of the hydrogen doublet sites with the rest of the hydrogen bonded to the silicon lattice is very weak. This fact is determined from measurements of the spin-lattice relaxation rates for the doublet, which are at least an order of magnitude slower that those of the bonded hydrogen below about 7 K [25]. If the doublet sites were closely coupled to the bonded hydrogen, then the rate would follow that of the bonded hydrogen due to rapid spin diffusion. The probable cause for the lack of coupling is that for some reason the doublet sites are physically removed from the bonded hydrogen. Because there are so few doublet sites, the separation may simply be a random occurrence. The temperature dependence of the spin-lattice relaxation rates is also much more rapid than that of the bonded hydrogen. The reason for this anomalous temperature dependence is not known.

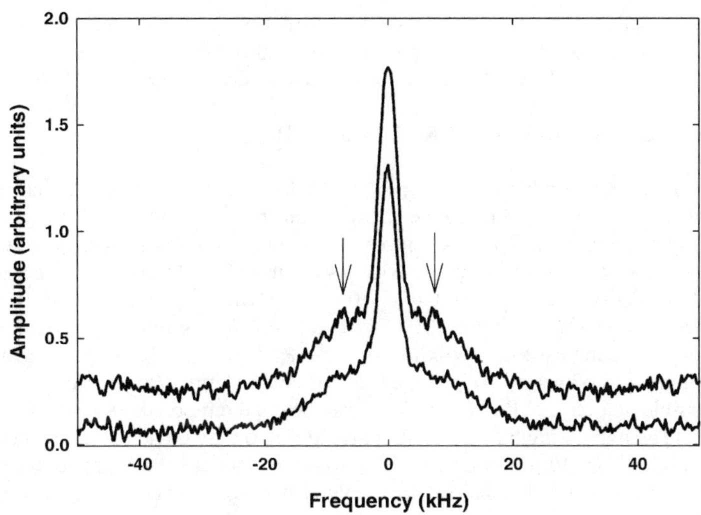

Figure 5. ^{1}H NMR echo lineshapes in a hydrogen-diluted sample of a-Si:H before (bottom trace) and after (top trace) irradiation with light. The arrows show the appearance of a paired hydrogen site that appears only after optical excitation. The top trace is displaced vertically for clarity. See text for details.

Because hydrogen has long been implicated in the Staebler-Wronski effect, many scenarios have been suggested to provide the metastable hydrogen complexes assumed to accompany the light induced production of metastable silicon dangling bonds [26]. We can rule out many of the specific suggestions. First, we can eliminate any models in which the sites do not involve paired hydrogen atoms. Zafar and Schiff [27] first suggested the presence of electrically inactive, paired hydrogen sites in a-Si:H, and Branz proposed the existence of paired hydrogen sites to explain the optically induced production of metastable silicon dangling bonds [28].

There remain several specific paired-hydrogen sites that are at present consistent with the data. Since the separation between the hydrogen atoms is consistent with SiH_2, or dihydride bonding, one cannot rule out this site as a possibility [29,30]. Several other specific sites have been proposed, including the analog in a-Si:H of the H_2^* defect that occurs in crystalline silicon [31], molecular hydrogen [32], and a surface dimer at a fully hydrogenated multivacancy [33]. We can rule out these specific examples because the hydrogen atoms are too far apart, too close together, or too close to bonded hydrogen atoms, respectively.

In addition to the dihydride site, several other specific sites are consistent with the hydrogen-hydrogen separation of 2.3 Å. One site proposed by van de Walle and Tuttle [32], albeit as an intermediate state and not as the stabilizing defect, is a five-fold-coordinated silicon atom with two Si-H bonds. In order for this site to be correct, the stability must somehow be enhanced in the amorphous environment over that which is the case in crystalline Si. Recently, Chadi has suggested two closely spaced Si-H bonds (H_2^{**}) formed by distortion of an H_2^* configuration [34]. It remains to be seen which, if any, of these models is correct.

STAEBLER-WRONSKI EFFECT IN a-Ge:H

In a-Ge:H, both metastable and stable (irreversible) increases in the spin density are observed by electron spin resonance (ESR) after irradiation with 1.17 eV light [35]. The inducing and annealing kinetics compare well to the well-studied behavior in hydrogenated amorphous silicon (a-Si:H). Figure 6 shows the initial ESR derivative spectrum of a-Ge:H (smallest intensity), the spectrum after 10 hours of irradiation at 2 W/cm^2 (largest intensity), and the spectrum after annealing at 170 C for one hour (middle intensity). After ten hours irradiation, the increase is approximately 100 % of the dark spin density. However after annealing, approximately 70% of the induced defects remain in the sample. These are irreversible defects that do not anneal out at elevated temperatures even after longer annealing times. [The narrow signal near 3390 gauss in Fig. 6 may be due to e′ centers, which presumably occur at the a-Ge:H/SiO_2 interface. The e′ center is an electron trapped in an sp^3 orbital at a silicon atom, which is bonded to three oxygen atoms.] The irradiation also produces changes in the e' center, presumably due to either hydrogen motion at the interface or charge trapping at the defect. [35].

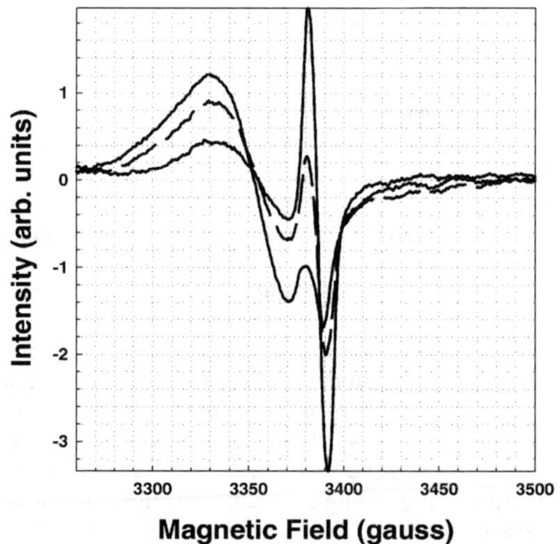

Figure 6. Derivative of the ESR absorption signals in a-Ge:H at 40 K. The least intense solid, most intense solid, and dashed lines are the initial dark signal, the signal after ten hours of irradiation at 300 K with 1 W/cm^2 of 1.17 eV light, and the signal after subsequent annealing at 171 C for one hour. All traces were run at 40 K to improve the signal-to-noise ratios. See text for details.

Figure 7 shows the growth of the total ESR signal in a stack of a-Ge:H films under irradiation with 0.5 and 1 W/cm^2. Both growth curves depend on time as $t^{1/3}$, which is similar to the kinetics observed in a-Si:H [29] and is representative of a broad distribution in the time constants.

The irreversible spin density appears to have essentially the same time dependence as the reversible spin density [35]. If this is indeed the case, then subtracting the irreversible spin density from the data points in Figure 7 will merely shift the graph downward on the log-log plot but will not change the general shape. Therefore, we conclude that the curves shown in Fig. 7 are reasonable representations of the growth of the SW defects in a-Ge:H, although the absolute spin densities are not correct because of the increasing density of stable defects with time. One can also estimate the fraction of defects that is irreversible. Near saturation of the optically induced component, this fraction depends super linearly on the generation rate, G, roughly as G^2. This power dependence indicates a two-photon process. The reversible SW defects, on the other hand, are well fit to a $G^{2/3}$ dependence, just as occurs in a-Si:H [29]. There is also a small irreversible component to a-Si:H after many cycles of light irradiation and annealing, but this effect has not been studied in detail [23,24].

The similarities in the kinetics for producing and annealing the dangling-bond ESR signals between a-Ge:H and a-Si:H provide strong evidence that the SW effect has been observed in a-Ge:H. However, it is the differences between these two materials that will be important in testing the various models proposed to explain the SW effect.

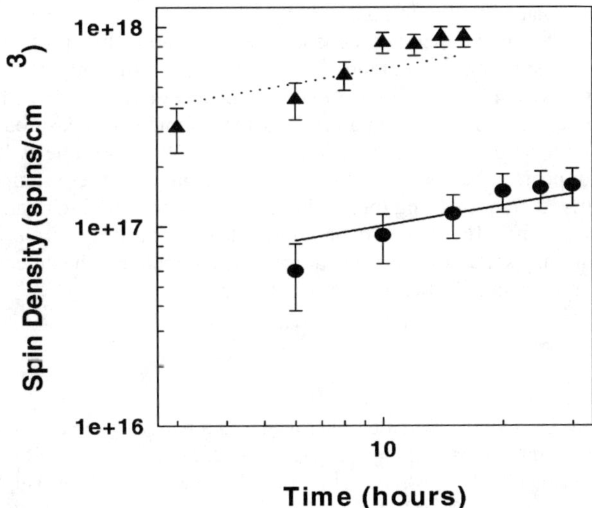

Figure 7. Growth of the total (dark and light induced) ESR absorption signals in a-Ge:H after irradiation with 0.5 W/cm^2 of 1.17 eV light (solid circles) and 2.0 W/cm^2 of 1.17eV light (solid squares). All measurements were taken at 40 K. The dotted and solid lines are power law fits to the data with an exponent of 1/3 ($t^{1/3}$). See text for details.

STAEBLER-WRONSKI EFFECT IN TRITIATED a-Si:H

The tritiated samples of a-Si:H (a-Si:H,T) were prepared in 1996 at the University of Toronto [36] using a DC glow discharge technique and tritium gas provided by a nuclear power plant. The sample shown in Fig. 8 was deposited at a substrate temperature of 225 C on a glass substrate. This sample was 0.26 microns thick and contained 1.092×10^{17} tritium atoms. High temperature, tritium effusion experiments determined an initial tritium concentration of approximately 6.0 at. % [36].

The a-Si:H,T sample was first measured after 7 years of uninterrupted tritium decay at 300 K. This 7-year spin density was orders of magnitude lower than would be expected if each tritium decay produced a silicon dangling bond. Also, the defects anneal in a manner that is very similar to the annealing of light induced defects in non-tritiated a-Si:H. As shown in Fig. 8, the saturated (7-year) spin density was 1.2×10^{17} spins/cm^3, and the annealed spin density was 2.7×10^{16} spins/cm^3.

Figure 8 shows the growth of the ESR signal at 300 K (circles) and the growth of the optical absorption at 1 eV as measured by photothermal deflection spectroscopy (PDS, triangles). Since the PDS is known to measure the density of silicon dangling bonds regardless of the charge state, these results show that, as in a-Si:H, the dangling bonds produced are essentially all neutral. The solid line that fits the data represents a stretched exponential rise to saturation with an exponent β that is very similar to that observed in a-Si:H (β=0.6 is the best fit). The dashed line through the data is a power law dependence on time ($t^{1/3}$) often taken to approximate the growth in a-Si:H. The solid line that rises rapidly is the predicted defect density if each tritium decay produced a metastable silicon dangling bond. Clearly there is some type of room temperature annealing going on.

To check for the temperature dependence of this annealing process the sample was stored at 77 K and warmed to 300K for at most 20 minutes before running at 77 K. These data are shown as the squares in Fig. 8. Finally, to eliminate any effects of short time annealing at 300 K on the low temperature growth, after annealing, the sample was measured at 40 K for30 hours. These data, which are not shown, indicate that the growth is the same at 40 K for up to 30 hours.

There appears to be some sort of athermal annealing in these samples. One possibility is that many atoms rearrange during the production of silicon dangling bonds as has been suggested by several experiments [16,17]. A second possibility is that the electrons emitted in the beta decay of tritium may somehow heat the lattice enough to anneal the defects even at 40 K. The correct explanation awaits further experiments.

SUMMARY

NMR measurements have demonstrated that a paired hydrogen site, where the hydrogen atoms are separated by approximately 2.3 Å apart, stabilizes the neutral silicon dangling bonds produced in the Staebler-Wronski effect in a-Si:H. ESR and PDS measurements in a-Ge:H have shown the presence of the Staebler-Wronski effect in this material, which will provide a second model system in which to study the Staebler-Wronski effect. Measurements in tritiated a-Si:H has shown the surprising result that most silicon dangling bonds created by tritium decay are annealed even at low temperatures. At present this result is not fully understood.

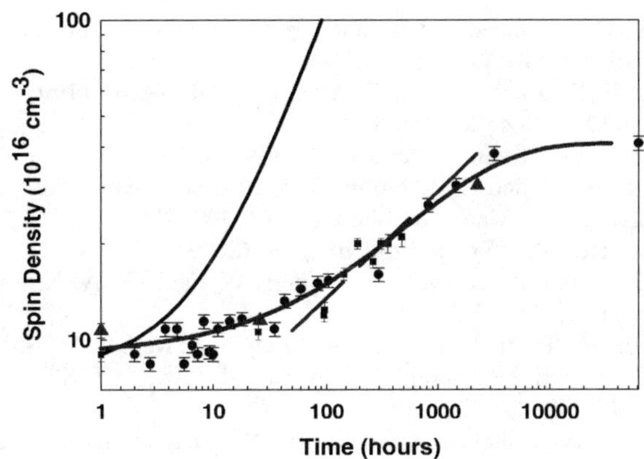

Figure 8. Growth of silicon dangling bonds in tritiated a-Si:H. Solid circles and solid squares show ESR data taken at 300 and 77 K, respectively. Solid triangles show 300 K optical absorption data at 1 eV and scaled to the ESR results. The solid and dashed lines through the data represent a stretched exponential rise to saturation and a power law fit, respectively. The solid line that rises rapidly is the curve expected if each tritium decay produced a metastable, neutral silicon dangling bond. See text for details.

ACKNOWLEDGEMENTS

The authors are very grateful to their co-authors who contributed to the work described in this review. In particular, we thank D. E. Carlson, G. Ganguly, M.M. de Lima, Jr., F.C. Marques, J. Viner, S. Zukotynski, E. Johnson, and P. Stradins for their essential contributions to the research. The work at the University of Utah was supported by NREL under contract #ADJ-2-30630-23 and NSF under grant #DMR-0073004.

REFERENCES

[1] C. G. van de Walle, Phys. Rev B **49**, 4579 (1994).
[2] K. J. Chang and D. J. Chadi, Phys. Rev. Lett. **62**, 937 (1989).
[3] S. B. Zhang and W. B. Jackson, Phys. Rev. B. **43**, 12142 (1991).
[4] D. Staebler and C. R. Wronski, Appl. Phys. Lett. **31**, 292 (1977).
[5] H. Dersch, J. Stuke and J. Beichler, Appl. Phys. Lett. **38**, 456 (1980).
[6] T. Su, P. C. Taylor, G. Ganguly, and D. E. Carlson, Phys. Rev. Lett. **89**, 015502-1 (2002).
[7] J. Whitaker and P. C. Taylor, unpublished.
[8] P.V. Santos, C. F. de O. Graeff, and I. Chambouleyron, J. Non-Cryst. Solids **128**, 243 (1991).

[9] C. F. de O. Graeff, P. V. Santos, G. Marcano, and I. Chambouleyron, Twenty First IEEE Photovoltaic Specialists Conf., v. 2, 1564 (1990).

[10] K. Eberhardt and G. H. Bauer, J. Non-Cryst. Solids **164-166**, 19 (1993).

[11] J. Lyou, J. Korean Phys. Soc. **28**, 594 (1995).

[12] T. Unold, Mat. Res. Soc. Symp. Proc. **336**, 287 (1994).

[13] H. Fritzsche, P. Stradins, and G. Belomoin, Mat. Res. Soc. Symp. Proc. **420**, 563 (1996).

[14] J. D. Cohen, Solar Energy Mat. and Solar Cells **78**, 399 (2003).

[15] J. D. Cohen, Mat. Res. Soc. Symp. Proc. **in**, press (2003).

[16] R. E. Norberg, J. Bodart, R. Corey, P. A. Fedders, W. Paul, W. A. Turner, D. Pang, and A. Wetzel, MRS Symp. Proc. **258**, 377 (1992).

[17] Y. Zhao, D. Zhang, G. Kong, G. Pan, and X. Liao, Phys. Rev. Lett. **74**, 558 (1995).

[18] S. Zukotynski, F. Gaspari, N. Kherani, T. Kosteski, K. Law, W.T. Shmayda, C.M. Tan, J. of Non-Cryst. Solids **299-302**, 476 (2002).

[19] J . Whitaker, J. Viner, S. Zukotynski, E. Johnson, P.C. Taylor, and P. Stradins, MRS Symp. Proc. (2004), in press.

[20] T. Su, S. Chen, P. C. Taylor, R. S. Crandall, and A. H. Mahan, Phys. Rev. B **62**, 12849 (2000).

[21] J. Jeener and P. Broekaert, Phys. Rev. **157**, 232 (1967).

[22] D. E. Carlson, K. Rajan, and D. Bradley, in *Proc. 26th IEEE Photovoltaic Specialists Conf.* (IEEE, New York, 1997), p. 595.

[23] S. Nitta, YU. Takahashi, and M. Noda, J. de Phys. **42**, C4-403 (1981).

[24] N. Schultz and P. C. Taylor, MRS Symp. Proc. **609**, A3.4.1 (2001).

[25] T. Su, P. C. Taylor, G. Ganguly, and D. E. Carlson, J. Non-Cryst. Solids (2004), in press.

[26] *Hydrogen in Crystalline Semiconductors*, S.J. Pearton, J.W. Corbett, M. Stavola, eds. (Springer-Verlag, Berlin, 1992); *Hydrogen in semiconductors II*, N. H. Nickel, ed. in Semiconductors and Semimetals, Vo. 61 (Academic Press, San Diego, 1999).

[27] S. Zafar and E. A. Schiff, Phys. Rev. B **40**, 5235 (1989).

[28] H. M. Branz, Solid State Commun. **105/6**, 387 (1998); Phys. Rev. B **59**, 5498 (1999).

[29] M. Stutzmann, W. Jackson, and T. T. Tsai, Phys. Rev. B **32**, 23 (1985).

[30] M. Stutzmann, private communication (2004).

[31] R. Biswas and Y. –P. Li, Phys. Rev. Lett. **82**, 2512 (1999).

[32] C. G. Van de Walle and B. Tuttle, MRS Symp. Proc. **557**, 275 (1999).

[33] S. B. Zhang and H. M. Branz, Phys. Rev. Lett. **87**, 105503 (2001).

[34] D. J. Chadi, Appl. Phys. Lett. **83**, 3710 (2003).

[35] J. Whitaker, M.M. de Lima Jr., F.C. Marques, and P.C. Taylor, J. Non-Cryst. Solids, (2004), in press.

[36] T. Kosteski, N. P. Kherani, P. Stradins, F. Gaspari, W. T. Shmayda, L. S. Sidhu, and S. Zukotynski, IEEE Proc.-Circuits Devices Syst. **150**, 274 (2003).

On the Mechanism of Hydrogen Diffusion in Si Solar Cells Using PECVD SiN:H

B.L. Sopori,[1] Y. Zhang,[2] R. Reedy,[1] K. M. Jones,[1] Y. Yan,[1] M. M. Al-Jassim,[1]
J. Kalejs,[3] and B. Bathey[3]
[1] National Renewable Energy Laboratory, 1617 Cole Blvd., Golden, CO 80401
[2] Bioarray Solutions, Warren, NJ 07059
[3] RWE Schott Solar, Billerica, MA 01821

ABSTRACT

A mechanism for transport and diffusion of H in a silicon solar cell by PECVD SiN:H process is proposed. Plasma-induced surface damage "stores" H during the nitride deposition, which is driven into the bulk of the solar cell during metal-contact firing. Theoretical and experimental results are given that verify this mechanism.

INTRODUCTION

Commercial Si solar cells are fabricated on low-cost wafers that contain high concentrations of impurities and defects. Most of these impurities and defects are removed or annihilated during solar cell fabrication by the gettering action associated with process steps such as phosphorous diffusion and Al alloying—processes used for formation of N/P junction and metallization contacts, respectively [1]. However, significant concentrations of impurities and defects continue to remain in the cell in electrically active states. Hydrogen has been used very successfully to passivate these residual defects and impurities, leading to improvements in cell performance. In some cases, cell efficiency can increase from 10% -11% to 14% – 15%.

Until recently, H passivation was done as a separate process step in which H was diffused at temperatures of 300°– 400°C by a plasma process. However, a new approach is now being used in the photovoltaic (PV) industry in which H passivation is a byproduct of a multifunctional process [2,3]. This process consists of deposition of a SiN:H film as an antireflection (AR) coating, and its subsequent processing to form the front metal contact by firing a screen-printed ink-pattern directly through this layer. Although this process is now used universally in the PV industry, a basic understanding of various mechanisms involved in it is severely lacking. To date, the process optimization is done empirically, for each material type or each batch of wafers, simply by optimizing the cell performance through trial and error. However, detailed studies are needed to understand each function of this process separately and to evaluate their synergistic effects. This understanding will help in optimum process design for higher cell efficiencies. One of the issues that need to be investigated is how H is transported from plasma into the bulk of the solar cell. In this paper, we describe the mechanism for H transport based on our theoretical model, and report experimental details for its verification.

SiN:H DEPOSITION AND SUBSEQUENT PROCESSING

In a typical commercial SiN:H-processing sequence, a thin layer (about 750 Å) of SiN:H is deposited on an N/P device by a plasma-enhanced chemical vapor deposition (PECVD) technique, using NH_3 and SiH_4 as process gases. The deposition is done in a temperature range of 300° - 400°C. The parameters of the deposition process are selected to attain the best antireflection effect and minimal optical absorption in the film. Next, a Ag-based contact-metallization pattern is screen-printed and then fired at about 800°C through the nitride using an RTP-like, conveyor-belt furnace. This RTP-like process is short, typically only about 10 s. In this step, the metal penetrates through the nitride to form a contact with the N/P device [4]. This process step is primarily optimized to produce a low-resistance ohmic contact, without shunting the junction. Minority-carrier lifetime (τ) studies have determined that the firing step is accompanied by a large increase in the τ due to passivation of impurities and defects. This passivation is attributed to the diffusion of H. FTIR spectroscopic studies have shown that the firing step also results in the release of H from SiN:H film. This evidence has prompted some researchers to assume that the evolved H (or a fraction of it) diffuses into Si and is responsible for passivation of impurities and defects. However, it is difficult to reconcile with this assumption because intuitively one would expect H to effuse out into the ambient rather than to go into Si. Indeed, from studies based on Si:H and SiN:H bond dynamics during RTP, it was concluded that nearly all evolved H escapes into the ambient [5-7]. This conclusion has prompted some authors to believe that such a process should not work. However, the fact remains that this process works very well, implying that a different mechanism must be involved.

PROPOSED MECHANISM OF H TRANSPORT AND DIFFUSION

We propose a new mechanism for transport of H and its diffusion into solar cell, during SiN:H processing, which invokes trapping/detrapping of H by process-induced defects [8,9]. Process-induced defects are generated by the damage produced at the surface of a wafer (or a solar cell) when it undergoes processes such as ion implantation or plasma treatments by energetic ions. Other steps, such as mechanical or chemical mechanical polishing, can also introduce surface damage [10]. We invoke this mechanism because, at low temperatures (below about 400°C), H diffusion in Si is strongly controlled by trapping and detrapping processes by impurities and defects. We believe that in a plasma nitridation process surface damage is caused during the initial stages of nitride deposition. The damaged surface has a very high "effective solubility" for H; it captures H from the ionized species in the plasma. Because the detrapping rate of H from the damage is quite low, the H is retained near the surface during the deposition.

During the firing step, the temperature of the wafer reaches about 800°C. The "stored" H is detrapped and, because of high diffusivity in the bulk of the cell at this temperature, it rapidly diffuses through the entire thickness of the solar cell. The diffused H is then responsible for passivation of impurities and defects. Here, we will illustrate through theoretical modeling that the above assumption of trapping and detrapping by the damage can account for diffusion of sufficient H into the bulk of the cell. Next, we will show experimental evidence of the damage, and trapping and release of H after various steps of the nitridation process.

The mechanism proposed here explains not only the mechanism of hydrogen transport and diffusion, but also many other observations in solar cell passivation. In particular, it explains a fact that good passivation of a solar cell is obtained only when a very specific sequence of conditions are followed, which includes an important step of pretreatment in a H plasma prior to initiation of deposition.

SUMMARY OF THEORETICAL MODELING

We have modeled hydrogen diffusion for a plasma process. Our modeling includes trapping and detrapping due to both the process-induced, as well as the bulk, defects. For the sake of completeness, we will first summarize salient results of our theory. Using the model of our previous publications (and references cited therein), a typical total process-induced trap (PIT) density may be described by the following expression:

$$[T_{tot}] = T_0 \exp\left[-\frac{x}{a+bt}\right] + T_b \quad ,$$

where the first time-dependent term is due to process damage and T_b is a constant bulk trap level. The details of this mathematical model are presented elsewhere [9]. Here, we show some results of these calculations to illustrate the main features of the H transport. We will use this trap profile to determine the H profile during the deposition of SiN:H. A typical deposition time for an optimized AR coating is about 6 minutes. We will use this duration for our calculation, realizing that the surface of the solar cell is not directly exposed to the plasma during the entire time of deposition. However, there are two reasons for using this value for an illustrative example. One, each SiN:H deposition follows a start-up recipe that includes ignition of a plasma without allowing nitride deposition to take place (i.e., SiH_4 is shut off). This initial time can be as much as a few minutes. Second, it is likely that as the nitride builds up, the surface damage continues to occur at a slower rate as the film thickness increases.

The calculated profile of H in a Si wafer during the nitride deposition is shown by the dotted line in Fig. 1.   Here, we have assumed that the bulk trap density in the wafer is 10^{17} cm^{-3}. Figure 1 also shows a magnified view of the near-surface profile (dotted line) in the inset. It is seen that H concentration drops very steeply near the surface, and the distribution immediate in the vicinity of the surface is primarily controlled by the PIT distribution. Likewise, the bulk trap density determines the distribution away from the surface. Next, the distribution of H after the 10-s RTP is shown by the solid line in Fig. 1. The solid line in the inset shows the magnified near-surface distribution of H. In this step, a part of the "stored" H is rendered free and diffuses deeply into the bulk of the cell or wafer.   It is important to recognize that the bulk traps can exercise a high degree of influence on this diffusion. Figure 1 shows that in both cases the trap density in the bulk (T_b) determines the bulk H concentration of the solar cell.

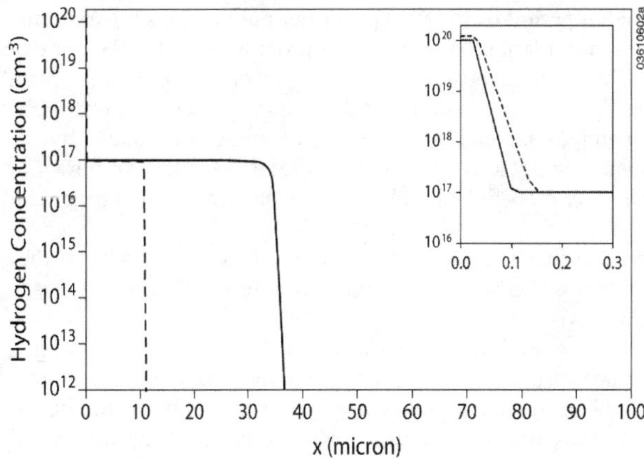

Figure 1. Calculated H profiles corresponding to a plasma exposure of 6 min (dotted) and after RTP (solid) processing at 800°C for 10 s. The inset shows the enlarged profile near the surface. The calculations have assumed a trap density of 10^{17} cm^{-3}.

The above calculations are aimed to show that even for a short exposure to plasma, a very high concentration of H can be introduced below the surface of Si. This process is similar to a pre-deposition in a dopant diffusion process used in the semiconductor device fabrication. These calculations were performed for a material of a very low quality, with a trap density $T_b = 10^{17}$ cm^{-3}. For more realistic material that yields solar cell efficiency of about 15%, the trap density has a value of ~10^{16} cm^{-3}. The diffusion behavior of hydrogen in this material can be different. Figure 2 shows similar plots of H profiles before and after RTP, for a trap density of 10^{16} cm^{-3}. It is seen that a reduced bulk trap density produces a deeper diffusion of H during the SiN:H deposition, and that H is distributed nearly uniformly in the entire cell after the RTP step.

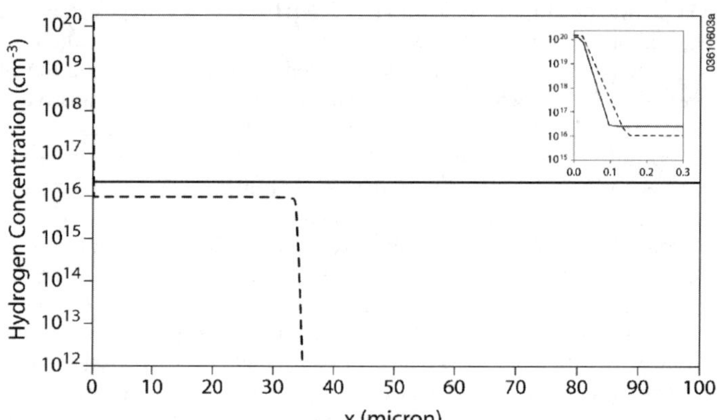

Figure 2. Calculated H profiles corresponding to a plasma exposure of 6 min (dotted) and after RTP processing at 800°C for 10 s (solid). The inset shows the enlarged profile near the surface. The calculations have assumed trap density of 10^{16} cm^{-3}.

EXPERIMENTAL

In our hypothesis, we have postulated the presence of a damaged surface layer to account for H storage. Formation of such damage is well known if the plasma has a high ion energy or processing is done for an extended period of time [11]. Earlier publications have discussed the nature of defects generated by ion implantation and plasma processing of H in Si for extended periods of time (15 – 30 min). These defects appear primarily as dislocation loops, platelets, and accumulation of H as "bubbles." However, in a nitride deposition process, the exposure to plasma is very short. Unfortunately, characteristics of defects/damage introduced by short exposure to low-energy H plasma are not known. Furthermore, it is not known how they interact with H, and how thermal processing modifies them. We have carried out experiments to investigate the following:

1. Is there a surface damage associated with the PECVD process? If so, what is the nature of damage? Is H "stored" below the Si surface, and what it its distribution?
2. How is the "stored" H redistributed upon firing?

It is noteworthy to mention that experimental verification of damage, and H storage associated with it, must be performed using samples that are representative of the commercial materials and their processing. Careful sample preparation and analysis is needed because these samples are multicrystalline and have rough surfaces even at submicron sampling for TEM observations. Furthermore, it is important that the plasma conditions and time should be same as those used for SiN:H deposition in solar cell processing. These requirements prevent use of some easier solutions. For example, although substitution of H-bearing gases by D-bearing gases would greatly improve the sensitivity for SIMS measurements, use of D-bearing gases changes the process conditions. Likewise, as shown in this paper, the nature of damage and defect generation is different in a short exposure to plasma than what could be concluded from defect generation due to higher energy and/or longer exposure to plasma.

We have investigated EFG material from RWE Schott Solar, and used all processing steps of SiN:H as used in the commercial solar cell fabrication. P-type EFG ribbons were processed to deposit SiN:H layers about 750 Å thick. One set of these samples was fired through the regular firing process (without screen printing). Another set of samples was processed through a P diffusion to have a junction prior to nitride deposition. Here, we will only present the results of samples without the junction.

Figure 3 is a high-resolution cross-sectional TEM image of an EFG ribbon after deposition of a SiN:H layer and before RTP step. The Si lattice image corresponding to (011) pole, and the nitride side is labeled. This image clearly shows the presence of defects in the Si region, which extends about 10–20 nm below the interface. The nature of damage appears to be in the form of clusters of point defects resulting from the low-energy ion bombardment. Several samples were examined with TEM, but they all confirmed the absence of any extended defects such as dislocations or platelets, which are typically seen in plasma-hydrogenated samples. One can also see that the interface between Si and SiN:H is rough. The roughness of the interface is intrinsic to the nature of the growth of these ribbons. The surface roughness makes it difficult to characterize near-surface defects using a standard bright-field image.   However, from Fig. 3, it is clear that PECVD nitridation creates damage at the Si surface. The depth of the damage ranges between 20 and 50 nm. Because our samples are multicrystalline, the process-induced damage appears to vary from grain to grain.

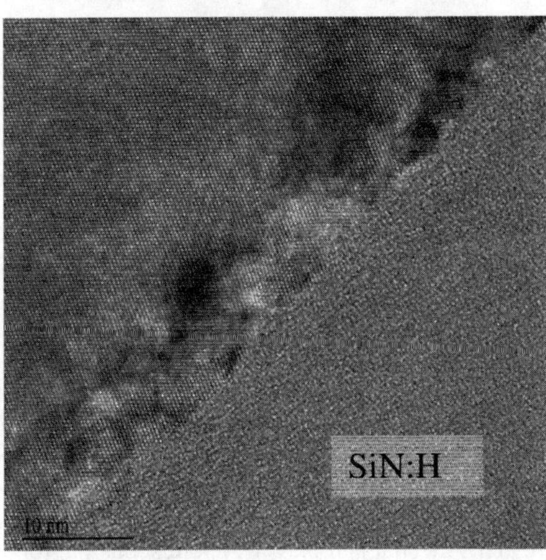

Figure 3. A [011] cross-sectional TEM lattice image of a Si ribbon deposited with a commercial SiN:H (before alloying step) showing surface damage.

Figure 4 (dotted line) shows a SIMS plot of H below the interface of a sample in which the SiN:H layer was removed after the deposition [12]. It is seen that a high concentration of H exists at the surface, and the profile is quite similar to the calculated profile shown in Fig. 2. Because of the poor sensitivity of SIMS, we cannot see the profile below the H concentration of 10^{18} cm^{-3}. Figure 4 also shows the H profile in a sample in which a SiN:H-coated sample was fired using the standard firing process for solar cell fabrication. The nitride layer was removed prior to SIMS measurements. It is clear from the figure that H has diffused into the Si wafer and that the surface concentration is lower after firing. The profile after firing differs somewhat from that shown in Fig. 4. However, the profile seems quite reasonable if we assume that some of the damage gets "healed" during the high-temperature firing (which is not taken into account in our calculations

We have also performed TEM investigations of fired samples. The majority of the samples show very little damage after firing. However, we have found regions of samples that have extended defects – that look very similar to plasma hydrogenation, but with a much smaller scale. Figure 5 shows a cross-sectional TEM of a fired sample with SiN:H. The defects below the nitride layer consist of platelets and "bubbles," as identified in the image. Thus, some of the damage in the form of point defect clusters due to nitridation, has developed into extended defects.

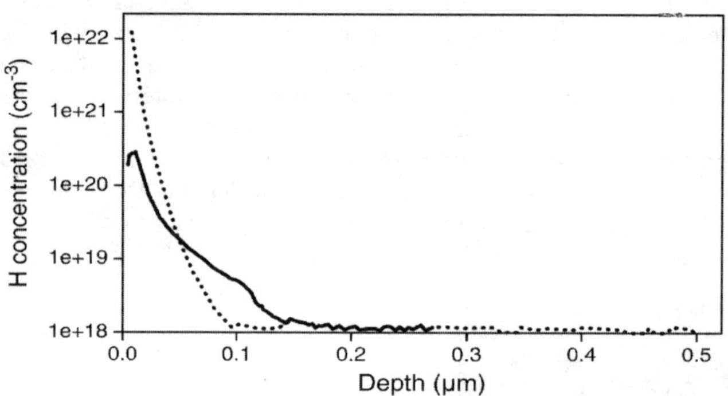

Figure 4. SIMS profiles of H before (dotted) and after firing (solid) the nitride-coated wafer. Nitride was removed prior to SIMS measurement.

DISCUSSION/CONCLUSION

We have shown that the proposed mechanism of hydrogen diffusion, mediated by process-induced defects, is responsible for the transport of H into the solar cell. Defects are induced at the surface by the nitridation process in the form of point-defect clusters, leading to a high solubility of H. A subsequent rapid anneal diffuses the H into the bulk of the Si, which passivates the residual impurities and defects. Some of the damage is healed by the thermal anneal, but some of the damage is converted into extended defects that are similar to plasma-induced defects (using plasma hydrogenation), but much smaller in size.

Existence of this mechanism explains why passivation is sensitive to how the nitridation process is performed and not to the composition of the nitride. It also suggests that nitride deposited by other techniques, such as reactive ion sputtering, which create a temporary damage layer near the surface, may also work well. This mechanism may also be responsible for enhanced passivation observed in Si wafers with a junction at the surface [13]. It is important that the firing process be designed to mend the damage, because a damaged surface will exhibit high surface recombination of carriers and lower device efficiency. It may be pointed out that

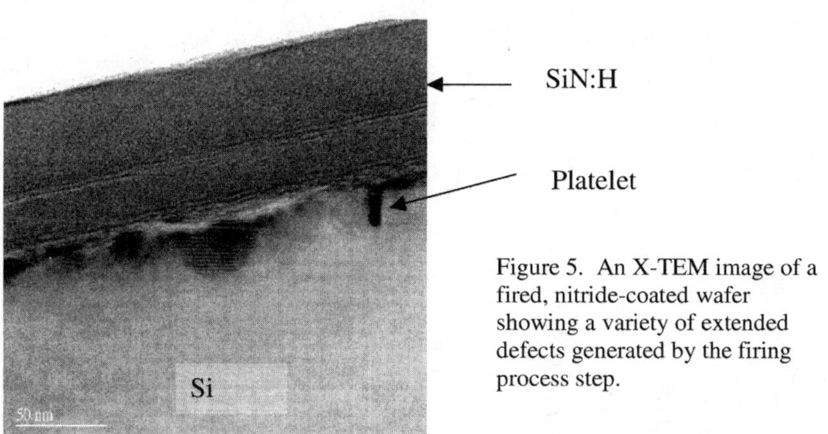

Figure 5. An X-TEM image of a fired, nitride-coated wafer showing a variety of extended defects generated by the firing process step.

the total amount of H introduced into Si during nitride deposition is very small. However, this is much larger than the amount of H needed to passivate minority-carrier lifetime-killing defects ($\sim10^{14}$ cm^{-3}) in the entire device.

REFERENCES

1. B. L. Sopori, J. Electronic Mat., **31**, 972 (2002).
2. V. Yelundur, A. Rohatgi, J-W. Jeong, A. M. Gabor, J.I. Hanoka, and R.L.Wallace, Procd. IEEE PVSC, 9 (2000).
3. J. Nijs et al., Proc. of the 1st World Conference on Photovoltaic Energy Conversion, Hawaii, 1242, (1994).
4. B.L. Sopori, J. Electronic Mat., **32**, 1034(2003).
5. W. M. Arnold Bik, R.N.H. Linssen, F.H.P.M. Habraken, and W. F. van der Weg, Appl. Phys. Lett., **56**, 2530 (1990).
6. Z. Lu, P. Santos-Filho, G. Stevens, M.J. Williams, and G. Lucovsky, J. Vac. Sci. Technol. A **13(3)**, 606(1995).
7. C. Boehme and G. Lucovsky, J. Appl. Phys., **88**, 6055 (2000).
8. B. L. Sopori, Y. Zhang, and R. Reedy, Procd. 29th IEEE PVSC, 222 (2002)
9. B.L. Sopori, Y. Zhang, R. Reedy, K. Jones, N.M. Ravindra, S. Rangan, and S. Ashok, Proc. MRS, **719**, 125(2002).
10. M. Symko, Hydrogen Diffusion Mechanisms in Silicon Solar Cells, Ph. D. Thesis, University of Denver, 1998.
11. B.L. Sopori, Y. Zhang, and N.M. Ravindra, J. Electronic Mat, **30**, 1616 (2001).
12. SiN:H layers were etched in dil. HF. Because of some concern if such a procedure will influence the existing H profile, we have done some detailed studies, which showed that the SIMS profiles were not influenced by etching. Further details of the influence of HF etching will be discussed elsewhere.
13. L. Mittelstadt, A. Metz, and R. Hezel, Solar Energy Materials and Solar Cells, **72**, 255(2002).

Hydrogen passivation of bulk defects and surface in silicon

Santo Martinuzzi and Olivier Palais
TECSEN Laboratory - University of Aix – Marseilles III –13397 France

ABSTRACT

Monocrystalline and multicrystalline silicon wafers were investigated with boron doping levels in the range 10^{15} cm^{-3} to 10^{16} cm^{-3}, respectively. Hydrogenation of the samples resulted from hydrogen-rich silicon nitride deposition by plasma enhanced chemical vapour deposition on the surface of the wafers. Passivation effects are observed after annealing and evaluated using minority carrier diffusion length (L) measurements, light beam induced current scan maps and lifetime (τ) measurements, by the contact-less phase shift technique. When applied at various excitation frequencies the phase shift technique leads evaluate the surface recombination velocity (S) and the actual bulk lifetime.
It was found, in multicrystalline silicon that isolated intragrain defects are well passivated while at grain boundaries and dislocation clusters deep levels are transformed into shallow levels. As a consequence L increases up to 80 % after firing the samples which the back surface was covered by a 2 µm thick aluminium layer. S decreases below 500 cm/s at the front surface covered by the hydrogen rich silicon nitride layer.

INTRODUCTION

Deep-level passivation by hydrogen in silicon has practical applications, although the exact passivation mechanism is not yet clearly understood. In multicrystalline silicon hydrogen diffuses easily in silicon via dislocations [4], although the diffusivity is strongly reduced in imperfect or in highly doped materials and it is emphasized that atomic hydrogen is the active species for defect and impurity passivation as well as dangling bond saturation [1; 2; 3].
Hydrogen is frequently introduced in silicon after deposition of a silicon nitride layer by plasma enhanced chemical vapor deposition (PECVD) in a hydrogen rich atmosphere [7]. Hydrogen migrates in the silicon wafer when the structure is fired at 600 - 700°C. This technique can passivate the surface and the bulk, and creates an antireflection coating for sunlight radiations. This is the reason why it is highly suited for solar cells [8].
The present paper deals with the hydrogen passivation of bulk and surface recombination centers by means of the PECVD deposition of a silicon nitride layer on monocristalline and large grain size polycrystalline wafers (multicrystalline silicon : mc-Si). The results show that the covered surfaces are well passivated, and that in mc-Si the recombination strength of grain boundaries (GBs) and dislocation clusters, are attenuated, due to the transformation of deep levels into shallow levels.

EXPERIMENTAL

Mc-Si wafers (so-called polix) made by Photowatt Int. S.A. that were p-type, boron doped to $2x10^{16}$ cm^{-3} were investigated. Their mean grain size was about 5 mm. GBs and dislocations are the main source of recombination centres. Companion wafers cut out from the same ingot and presenting the same kind and density of defects were used to make N$^+$P junctions by phosphorus diffusion from POCl$_3$ at 850°C for 20 min. One series was coated with a TiOx antireflection layer (TiO samples) and another one with a hydrogen rich silicon nitride layer (SiNH samples). TiO$_x$ was deposited a 450°C by atmospheric pressure CVD, while SiNH was obtained by a direct RF plasma reactor, with SiH$_4$ and NH$_3$ as precursors, at a temperature of 350°C.
Metallic contacts were obtained on both series by metal deposition and firing at 750°C. During the firing H atoms diffused into the silicon wafer.

Minority carrier diffusion lengths (L) were measured by means of the surface photovoltage technique while LBIC scan maps revealed the presence and the features of recombining defects. The collecting structures used for the LBIC scans were the N$^+$P junctions. Details of the technique have already been published [9].

Using a microcryostat the LBIC scan maps were also obtained at temperatures between 300 and 90 K. The evolution of the LBIC contrast (C) with T gives an indication about the depth of the energy level of the recombination centers : when T decreases and C increases shallow levels can be ascribed to the defects. Conversely when T and C decrease together, deep levels are concerned. This is due to the variation of the recombination lifetime (τ) with T (C is proportional to τ^{-1}) [10].

Effective lifetime of minority carriers were evaluated by means of the microwave contactless phase shift technique. This technique uses a weakly sine-modulated near-infrared exciting light and a microwave probe. The phase shift between the excitation and the power of the reflected microwaves depends on the modulation frequency of the exciting light. When carried out at two different light excitation frequencies, the lifetime measurements determine both the actual bulk lifetime τ_b and an accurate value of the surface recombination velocity S. Details of the method have already been given in ref. 11. The technique was applied to evaluate the passivation of the surface in samples covered by the SiNH layer. For comparison, investigations were also carried out on Czochralski grown monocrystals, both face polished, doped to 10^{15} cm^{-3} (N type or P type) which the surfaces were passivated by immersion in an iodine polyvidone solution.

RESULTS

The influence of hydrogen in-diffusion is easily verified by comparing the mean values of L measured in TiO and SiNH samples. It was found that L increases by at least 80 % after hydrogenation (i.e. from 140 to 260 µm).

The LBIC scan maps of companion TiO and SiNH samples show that, at room temperature, there is a marked contrast at GBs and dislocations lineages or clusters in TiO samples, while this contrast is very weak in SiNH samples. When the temperature T decreases in SiNH samples, C increases at GBs and dislocation lineages or clusters but not in homogeneous grains

The Fig. 1 scan lines which cross 3 GBs of a SiNH sample illustrate clearly the increase of C at these defects when T decreases from 300 K to 100 K. In TiO samples, C behaves differently, it decreases with T and practically disappears at GBs, which is well illustrated by the Fig. 2 scan lines which cross 2 GBs.

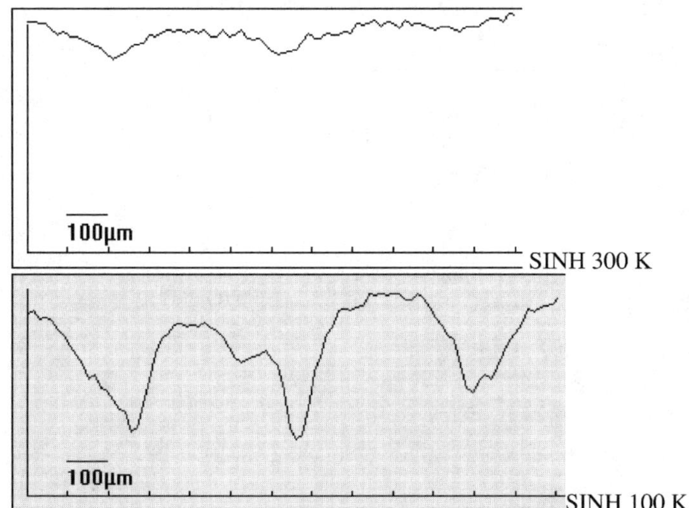

Figure 1. Photocurrent scan line of 3 GBs in SiNH samples which clearly show the increase of the contrast at these defects when the temperature decreases from 300 to 100 K and indicate that shallow energy levels control the recombination strength at the defects.

The increase of the contrast at GBs and dislocation clusters in SiN-H samples when T decreases suggests that the shallow levels observed in SiN-H samples are located at approximately 0.1 eV above the valence band. In non passivated samples, deep levels were reported to be close to the mid-gap [1; 2; 3].

Lifetime measurements confirm the improvement of the material bulk and give information about the surface passivation. The validity of the technique is verified by the investigation of monocrystals, in which it is found that S decreases from 5×10^3 cm/s to less than 80 cm/s, when the surfaces are passivated by immersion in the iodine solution, while the value of the bulk lifetime is approximately the same in passivated and raw samples. Notice that the use of the conventional microwave photoconductivity decay lead to τ_b only when the S value is known, but cannot give directly both the values of S and τ_b like the phase shift technique.

The evaluations of S in raw and passivated mc-Si wafers lead to the following results:

After immersion in the iodine solution, S decreases also from 7×10^3 cm/s to about 100 cm/s, however when SiNH layers cover the surfaces, the values of S are below 500 cm/s and depend strongly on the deposition conditions of the nitride layer. It appears that the deposition conditions which give the higher improvement of lifetime are not those which produce the stronger reduction of S, in agreement with ref. 12.

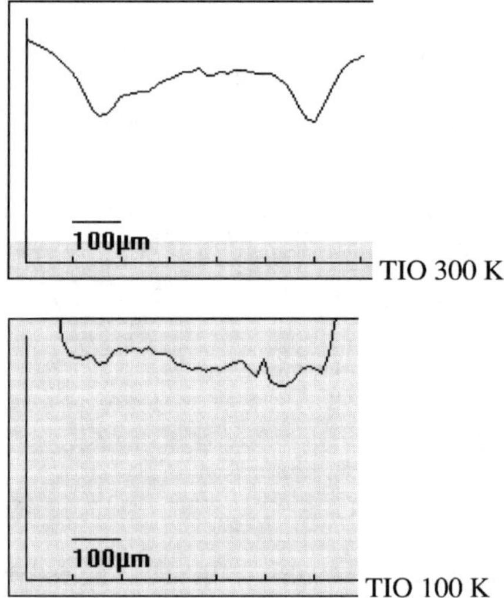

Figure 2. Photocurrent scan lines of two GBs in TiO samples which show that there is no contrast at GBs at 100 K and indicate that deep energy levels only control the recombination strength of the defects.

DISCUSSION

The results indicate that hydrogenation of the wafers by in-diffusion from the SiNH layer increases the minority carrier diffusion length. This improvement is due to a marked reduction of the recombination strength at any kind of imperfections. Table I indicates some species and defects which interact with hydrogen [1]. Such interactions explain the observed improvements. However, the evolutions of the LBIC contrast with temperature show that there is rather a transformation of deep levels into shallow levels due to an interaction of hydrogen with these defects. The presence of a nearby hydrogen species may reduce the nature of the electrical activity of the defects, and it should be emphasized that H atoms interact with deep levels in the band gap rendering them less active. Such an interaction may include either the formation of a stable impurity-hydrogen bond, or the rearrangement of the defect structure.

Notice that the transformation of deep levels into shallow ones is frequently observed after impurity-impurity interaction, one well known example is given by the FeB pairs in p-type silicon : dissolved interstitial Fe_i gives rise to a deep level (E_v + 0,39 eV) while the level of FeB pairs is at 0.10 eV above the valence band.

Table I. Impurities and defects which the recombination strength can be neutralized by interaction with hydrogen (1).

Impurities	Trapped Carrier	Energy Level in the gap (eV)
Cr	Electrons	0.31
Cu	Holes	0.2 ; 0.35 ; 0.53
Ni	Holes	0.18 ; 0.21 ; 0.33
Fe	Holes/electrons	0.32 ; 0.39
O-V	Electrons	0.18
Defects		
V-V	Electrons	0.22
Dislocations		Band of deep levels
Grain Boundaries		Bands of deep levels
µ-precipitates	Electrons/holes	

The different passivation efficiency of bulk and surface imperfections may be explained by the reduced passivation effect of the PECVD silicon nitride layer, which is deposited at low temperature. Although this layer contains hydrogen, it is less effective than pure silicon nitride layer deposited at high temperatures, i.e. above 800°C [12].

CONCLUSION

In this paper we show that hydrogenation of disordered silicon wafers, such as multicrystalline wafers,through the deposition of a SiN-H layer is a powerful means to increase minority carrier diffusion lengths. The SiN-H layer is a reservoir which feeds the silicon with hydrogen during subsequent annealings. Individual dislocations and dissolved impurities are well passivated in homogeneous grains, while grain boundaries and dislocation lineages or tangles are not totally neutralized. This is due to the fact that deep levels associated with individual defects are strongly passivated whereas deep levels are transformed into shallow ones at grain boundaries and dislocation clusters. A marked interaction of crystallographic defects and impurities may account for such behaviours.

ACKNOWLEDGMENTS

The author would like to thank PHOTOWATT Int. S.A. Bourgoin Jallieu – France for supplying companion sample.
This work was supported by CNRS/ECODEVE and ADEME - France.

REFERENCES

1. S.J. Pearton, J.W. Corbett and M. Stavola, in *Hydrogen in CrystallineSemiconductors*, edited by H.J. Queisser, Springer Series in Materials Sciences 16, Berlin, 1992.
2. J.I. Pankove and N.M. Johnson, in *Hydrogen in Semiconductors*, ed. by R.K.Willardson and A.C. Beer, Academic Press Inc., New York, 1991.
3. Defects in Silicon : Hydrogen, ed. by J. Weber and A. Mesli, European MaterialsResearch Soc., Symposia proceedings 76, Elsevier Amsterdam, 1999.
4. C. Dube and J.J. Hanoka, *Appl. Phys. Lett.* 45, 1135 (1984).
5. J.I. Pankove, in *Semiconductors and Semimetals*, ed. by Acad. Press Inc. New York, 261,1984.
6. J.M.E. Harper, J. Cuomo and H.R. Kaufman, *Ann. Rev. Mater. Sci.* 13, 413 (1983).
7. R. Hezel and K. Jäger, *J. Electrochem. Soc.* 59, 457 (1986).
8. R. Einhaus, F. Duerinckx, J. Szlufcik, F. Durand and S. Martinuzzi, *Mat. Science Engineering* B58, 81 (1999).
9. M. Stemmer, *Appl. Surf. Science* 63, 213 (1993).
10. I. Périchaud, Scott Mc Hugo and S. Martinuzzi, *Solid State Phenomena* 63, 53 (1998).
11. O. Palais and A. Arcari, *J. of Appl. Phys.*, 93, 8, 4686 (2003).
12. M.J. Kerr and A. Cuevas, *Proc. of 20 [th] IEEE Photovoltaic Specialists Conference*, p.102, 2002.

Formation of deuterium-related shallow donors in boron-doped diamond.

Jacques Chevallier[1], Zéphirin Teukam[1], Cécile Saguy[2], Rafi Kalish[2], Catherine Cytermann[2], Francis Bailly[1], François Jomard[1], Thierry Kociniewski[1], Dominique Ballutaud[1], Michel Barbé[1], James E.Butler[3], Céline Baron[4] and Alain Deneuville[4].

[1] Laboratoire de Physique des Solides et de Cristallogénèse, UMR CNRS 8635, 1 place A.Briand, 92195 Meudon Cedex, France

[2] Physics Department and Solid State Institute, Technion, Haifa 32000-Israel

[3] Naval Research Laboratory, Code 6174, Washington, DC 20375, USA

[4] Laboratoire d'Etudes des Propriétés Electroniques des Solides, CNRS B.P.166, 38042 Grenoble Cedex 09, France

ABSTRACT

It is well known that diffusion of deuterium in boron-doped diamond results in the passivation of boron acceptors with the formation of (B,D) complexes. In this work, we show that deuteration of boron-doped diamond can induce a p-type to n-type conversion under certain conditions. The n-type conductivity is governed by the ionization of shallow donors with a ionization energy of 0.34 eV. This is well below the lowest ionization energy of donors found up to now in diamond (0.6 eV for phosphorus donors). The electrical conductivity and the electron mobility can be as high as 6 S/cm and 430 cm^2/Vs at 300 K. The reversibility of the effect under thermal annealing and the necessity of excess deuterium to trigger the n-type conductivity suggest that deuterium is involved in the formation of the shallow donors. The present status concerning the understanding of their origin is discussed. In addition, we have found that, contrary to previous conclusions, deuterium can diffuse in type Ib diamond. The conditions where this diffusion is observed are presented.

INTRODUCTION

Diamond homoepitaxial layers grown by microwave plasma chemical vapor deposition (MPCVD) use a hydrogen-rich gas mixture of CH_4 and H_2 with more than 95% of hydrogen since atomic hydrogen is essential for the diamond deposition [1]. Owing to the abundance of hydrogen in the diamond growth system, it is necessary to consider the possibility of hydrogen incorporation, to investigate its influence on the electronic properties of diamond and, if any, the possibility of hydrogen incorporation during the processing steps for the fabrication of diamond-based electronic and optoelectronic devices.

Hydrogen at the surface of diamond is directly responsible for the negative electron affinity of (100) and (111) diamond surfaces [2,3] and is at the origin of its surface p-type conductivity. This effect is the result of an electronic charge transfer from diamond to surface adsorbates [4]. It has been used for the fabrication of high frequency diamond-based field effect transistors [5].

Theoretical calculations have shown that hydrogen is a positive-U character impurity in diamond [6,7]. Hydrogen atoms have a donor level at E_c-3 eV and an acceptor level at E_c-2 eV [6]. As in silicon, they act as counter-dopants behaving as donors in p-type diamond and as acceptors in n-type diamond. The deep donor character of hydrogen in diamond has also been established experimentally in muon experiments. In high resistivity diamond, neutral $Mu^o{}_{BC}$

are stable at temperatures as high as 800°C [8]. This suggests that the (+/o) donor level of hydrogen is deep enough to prevent the formation of H^+ at such high temperatures. For comparison, in high resistivity silicon, Mu^O_{BC} ionizes into Mu^+_{BC} at a temperature of about 200 K as a result of the ionization of the Mu_{BC} donor level (+/o) located at E_c-0.2 eV [9,10].

In boron-doped diamond, hydrogen atoms diffuse as protons [11,12,13]. However, their migration energy E_m is lower in diamond (0.1-0.2 eV) than in silicon (0.5 eV) [14,15,16]. Hydrogen diffusion is limited by trapping on boron acceptors [11]. The formation of (B,H) complexes results in the passivation of boron acceptors [11,17]. Their dissociation energy E_{diss} is 2.5 eV (1.2 eV in silicon) [18,19,20]. This indicates that the binding energy E_b (E_b=E_{diss}-E_m) of (B,H) pairs is much larger in diamond (2.3-2.4 eV) than in silicon (0.7 eV).

Hydrogen diffusion has also been investigated in n-type diamond. Exposure of phosphorus-doped or nitrogen doped diamond to a deuterium plasma introduces a deuterium accumulation in the near surface region but, up to now, there is no indication of deuterium diffusion deeper in the bulk even at temperatures as high as 1000°C [21].

In this work, we will show that the deuteration of boron-doped diamond not only lead to the formation of (B,D) complexes but can also yield, for specific boron-doped homoepitaxial layers, deuterium-related shallow donors. Fabrication of n-type diamond with satisfactory electrical properties is a challenge by itself. Up to now, phosphorus atoms have been the best n-type dopants in diamond with however a relatively high ionization energy (0.6 eV) and a limited solubility (about $3x10^{19}$ cm^{-3} in (111) homoepitaxial layers) [22,23]. Combined with low mobilities associated with the presence of growth defects in (111) epitaxial layers, this explains why the conductivities of phosphorus-doped diamond are limited to 10^{-3}-10^{-5} S/cm at 300 K. These conductivities are not satisfactory for electronic applications. We will show that our results open new possibilities in the research of shallow donors and the fabrication of n-type diamond with high electrical conductivities. Finally, we will report that, under some particular conditions, diffusion of deuterium can be clearly detected in nitrogen-doped diamond.

EXPERIMENTAL DETAILS

A set of homoepitaxial layers of boron-doped diamond was grown by microwave plasma chemical vapor deposition. Boron doping was achieved by the addition of diborane to the gas mixture. The growth conditions have been described elsewhere [24]. Their thickness and boron concentration were in the range 0.4 - 2 μm and $2x10^{17}$ – $4x10^{19}$ cm^{-3}. The epilayers were exposed to a deuterium plasma in conditions where deuterium diffuses in the whole thickness. The diffusion temperatures and times were chosen such as to ensure (B,D) complex formation (temperatures equal to or less than 550°C).

Secondary ion mass spectrometry (SIMS) analysis was performed with a CAMECA IMS 4f to determine the boron and the deuterium concentrations in the epilayers and to detect other possible contaminants such as oxygen or nitrogen. Electrical conductivity and Hall effect measurements were performed from 200 K to 900 K in the van der Pauw configuration after oxidation of the sample surface. They were performed in the as-grown state, the as-deuterated state and finally after deuteration followed by a thermal annealing.

RESULTS AND DISCUSSION

The p-type to n-type conversion of diamond upon deuteration

Figure 1 presents the boron concentration profile of an homoepitaxial diamond layer together with the deuterium concentration profile after a 550°C, 8 h exposure to a deuterium microwave plasma. The deuterium concentration is close to the boron concentration as a result of the (B,D) pair formation.

Figures 2a and 2b show the temperature dependence of the free carrier concentration and the carrier mobility after deuteration at 550°C, 8 h of this boron-doped epilayer. In figure 2a, state 1 corresponds to the p-type as-grown state ($p=2.5 \times 10^{19}$ cm^{-3}, $\mu_p=2.5$ cm^2/Vs). After deuteration (state 2), the sample converts into an n-type state ($n=2 \times 10^{19}$ cm^{-3}, $\mu_n=2$ cm^2/Vs). Under annealing at 520°C, 0.5 h (state 3) and then 520°C, 1 h (state 4), it remains n-type. Fitting the temperature dependence of the free electron concentration with the equation describing n(T) in a partially compensated n-type semiconductor [25], the donor activation energies are 0.23 eV and 0.31 eV for states 3 and 4 respectively. In state 4, the electron concentration is 1×10^{16} cm^{-3} and the electron mobility reaches 430 cm^2/Vs at 300 K. The mobility values and the activation energies in states 3 and 4 indicate that, in these states, the conduction is controlled by the ionization of the donors towards the conduction band. The n-type electrical conductivities in states 2, 3 and 4 are respectively 6.5, 2 and 0.5 S/cm at 300 K.

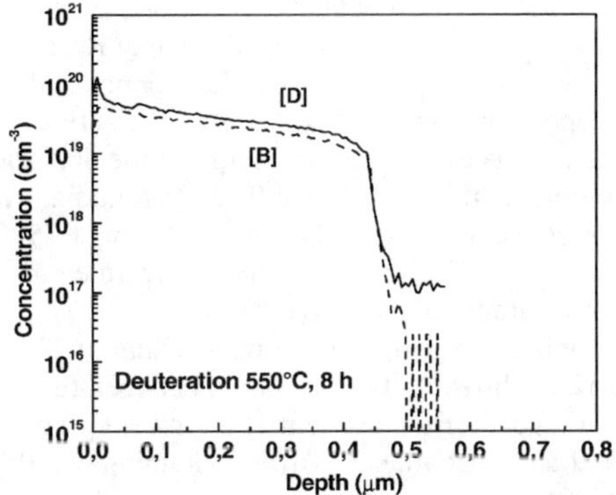

Figure 1. Deuterium profile in a boron-doped diamond epilayer after deuteration at 550°C, 8 h.

Under annealing at 650°C and above (states 5 and 6 of figure 2), the sample returns to the original p-type conductivity. The reversibility of the effect strongly supports a model where deuterium-induced shallow donors have been formed under deuteration. These donors dissociate around 520°C while (B,D) pairs remain since these pairs are stable up to 550°C [26,27]. At 600°C and above, (B,D) pairs dissociate. SIMS and out-diffusion experiments show that deuterium leaves the sample at 750°C and above [24,26].

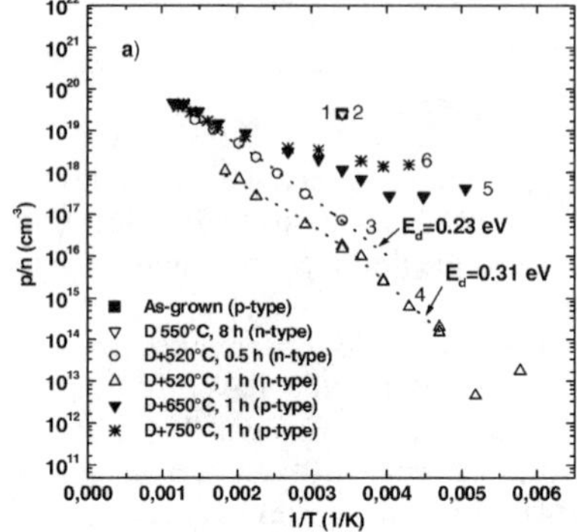
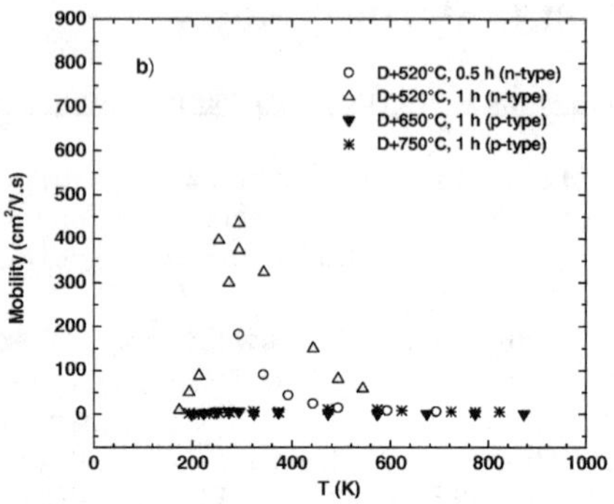

Figure 2. Temperature dependence of (a) the carrier concentration of a boron-doped epilayer ($[B]=1\text{-}4\times10^{19}$ cm^{-3}) in the as-grown (1), as-deuterated (2), deuterated and annealed at 520°C for 0.5 h (3) and for 1 h (4), annealed at 650°C for 1 h (5) and then 750°C, 1 h (6) states, (b) the mobility in deuterated and annealed at 520°C for 0.5 h and for 1 h, annealed at 650°C for 1 h and then 750°C, 1 h states. The dashed lines are obtained by fitting the data for states 3 and 4 to the equation describing n(T) in a partially compensated n-type semiconductor [25].

Similar results of p-type to n-type conversion have been obtained in samples containing lower boron concentrations and coming from two different laboratories: the Naval Research Laboratory in Washington and the National Institute of Advanced Industrial Science and Technology in Tsukuba. In the lowest boron-doped epilayers ($2\text{-}3\times10^{17}$ cm^{-3}), the activation energy of the donors is 0.34 eV which is considered as the ionization energy of the deuterium-related shallow donors [28]. The lower activation energies of 0.23 and 0.31 eV found above are due to the presence of higher shallow donor concentrations in the annealed sample (520°C, 0.5 h and then 1 h) discussed in figure 2. Infrared transmission spectra have confirmed that, in n-type converted samples, neutral boron acceptors are totally absent [24].

An important issue is the formation mechanism of the deuterium-related shallow donors. Answer to this question can be partly provided by following the correlation between the appearance of the n-type conductivity and the total deuterium concentration in the epilayer.

Table I presents the electrical properties of a 0.9 μm thick epilayer containing 3×10^{19} cm^{-3} boron in different states. In the as-grown state, the free hole concentration is 3.4×10^{16} cm^{-3} and the hole mobility is 31 cm^2/Vs. A first deuteration at 550°C, 8 h ensures deuterium diffusion in the whole epilayer. Careful SIMS analysis show a ratio [D]/[B]=1. In these deuteration conditions, we observe a decrease of the hole concentration and an increase of the hole mobility (table I). This is representative of a conventional passivation effect of the acceptors. After a second deuteration at 475°C, 12 h of the same sample, the deuterium concentration exceeds the boron concentration with a [D]/[B] ratio decreasing from 3 at the surface to 1 at the interface with the substrate. In addition, we observe the p-type to n-type conversion. These results suggest a correlation between the conversion and the presence of excess deuterium.

	As-grown	1st deuteration 550°C, 8 h	2nd deuteration 475°C, 12 h
type	p	p	n
carrier conc. (cm^{-3})	3.4×10^{16}	1.1×10^{15}	1.5×10^{12}
mobility (cm^2/Vs)	31	230	150
[D]/[B]	0	1	1-3

Table I. Evolution of the carrier concentration, carrier mobility and [D]/[B] ratio for different states of a 0.9 µm thick boron-doped epilayer containing 3×10^{19} cm^{-3} boron.

Interesting information have been reached from the evolution of the deuterium profile between the first and the second deuteration of another boron-doped sample ([B]=2-7x10^{19} cm^{-3}, p(300 K)=6x10^{16} cm^{-3}, μ_p=23 cm^2/Vs). As shown in figure 3a, the [D]/[B] ratio is equal to 1 after the first deuteration at 550°C, 8 h. This corresponds to a passivation of the epilayer since p=4x10^{14} cm^{-3} , μ_p=475 cm^2/Vs after this first deuteration. An additional deuteration at 475°C, 2 h creates an excess deuterium in the first half of the epilayer with [D]/[B]≈2. Beyond this region, one finds again [D]/[B]=1. According to the results of table 1, we believe that this excess deuterium is responsible for the conversion to n-type conductivity.

Now, let us compare the deuterium diffusion coefficients in the first and in the second deuteration. In the first deuteration (550°C, 8 h), deuterium diffuses in an epilayer where initially all boron traps are empty (no (B,D) complexes) (figure 3a). At 550°C, the deuterium diffusion is entirely controlled by a trapping-detrapping mechanism. The effective diffusion coefficient at this temperature can be evaluated from our previous investigations where we established the evolution of the (B,D) pair frequency dissociation ν versus temperature given by [18]:

$$\nu = 4\pi R_c D_{eff}[B_{total}] = \nu_0 \exp{-\frac{E_{diss}}{kT}} \tag{1}$$

where ν_0 is the attempt frequency, R_c the capture radius, $[B_{total}]$ the boron concentration and E_{diss} the dissociation energy of (B,D) pairs.

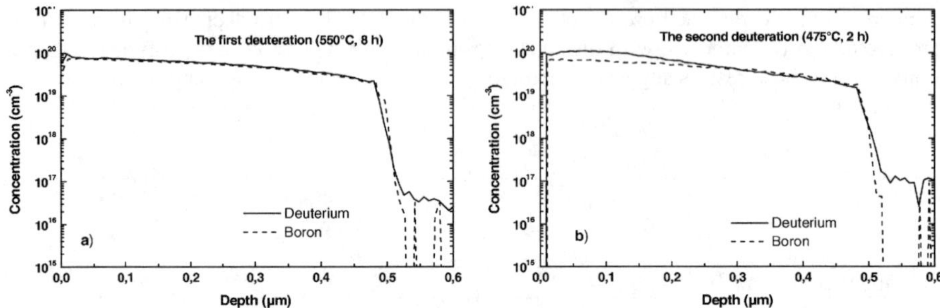

Figure 3. Deuterium diffusion profiles in a boron-doped epilayer containing 2-7x10^{19} cm^{-3} boron after a) the first deuteration at 550°C, 8 h, b) the second deuteration at 475°C, 2 h.

We conclude that, in the first deuteration at 550°C, the effective deuterium diffusion coefficient is 1.8x10^{-15} cm^2/s for [B]=5x10^{19} cm^{-3}. It drops down to 8x10^{-17} cm^2/s at 475°C. In this case, the deuterium diffuses slowly in spite of its high migration energy because its residence time on boron acceptors is much higher than the transit time from one boron to the next boron.

Before the second deuteration, we are in a initial situation where all boron are in the form of (B,D) pairs. As seen in figure 3b, this second deuteration at 475°C, 2 h generates an excess deuterium which is indicative that another deuterium-related complex is forming as a result of the interaction between deuterium and some species. However, this new interaction is weaker than the interaction between B$^-$ and D$^+$. Fitting the excess deuterium concentration [D$_{total}$]-[B,D] versus depth with a complementary error function (figure 4), one deduces a diffusion coefficient of 1.5x10^{-14} cm^2/s at 475°C for the excess deuterium.

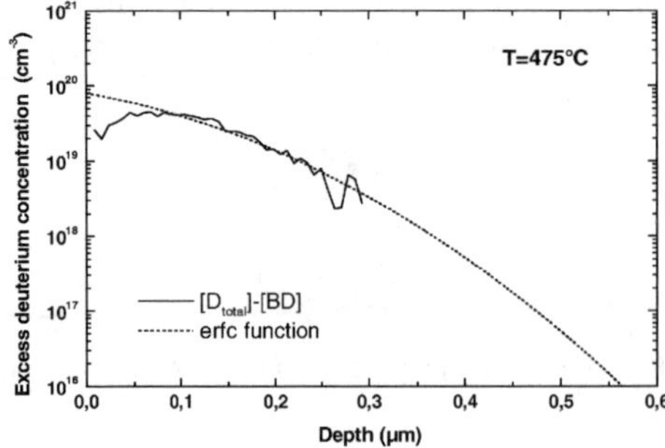

Figure 4. Excess deuterium diffusion profile in the same sample as in figure 3. This excess deuterium is [D$_{total}$]-[B,D]. Fitting the profile with an erfc function gives an effective diffusion coefficient of excess deuterium of 1.5x10^{-14} cm^2/s at 475°C.

This diffusion coefficient is 200 times larger than the diffusion coefficient of deuterium in the boron trap case (first deuteration). Consequently, deuterium diffuses much more freely in the saturated boron trap situation than in the boron trap situation.

Assuming that the presence of excess deuterium is responsible for the creation of shallow donors, we conclude that the shallow donor formation is the result of a two-step process. The first step is the passivation of all boron acceptors present in the epilayer. The second step is the formation of the new donors by trapping of deuterium on some species.

Presently, the nature of these species is unknown. A possibility would be the trapping of deuterium on particular defects giving rise to specific (D,defect) complexes in addition to the pre-existing (B,D) pairs. This is questionable since it would imply a high concentration of defects unless the defect itself is boron-related. Another possibility is the formation of boron and deuterium complexes with more than one deuterium per boron like (B,D_2) complexes. This is the model we initially proposed [24].

Several groups have performed calculations on (B,H_2) complexes in diamond. Goss et al concluded that (B,H_2) complexes are deep donors but they did not find any stable configuration in which boron can bind more than one hydrogen [29]. Dai et al found that (B,H_2) and (B,H_3) complexes have donor levels at 1.2 and 0.7 eV below the conduction band of diamond respectively but they did not report information about the absolute stability of the configurations they considered [30]. Surprisingly, they concluded that (B,H) complexes behave as shallow acceptors, even shallower than isolated boron, while it is known that hydrogen passivates boron acceptors. This contradiction with well established experimental results makes questionable the conclusions of Dai's paper.

It is important to note that the quality of the epitaxial layer is a key parameter for the success of the p-type to n-type conversion. An example of no conversion is shown in figure 5. This sample was doped with 4×10^{18} cm^{-3} boron and its free hole concentration and hole mobility were respectively $p = 1 \times 10^{15}$ cm^{-3} and $\mu_p = 550$ cm^2/Vs. The result of a deuteration at 550°C, 2 h was a significant passivation of the boron acceptors ($p = 2 \times 10^{14}$ cm^{-3} and $\mu_p = 760$ cm^2/Vs) in spite of the creation of an excess deuterium in the epilayer. We note a very large accumulation of deuterium at the epilayer/substrate interface. This accumulation is attributed to the trapping of deuterium on structural defects present at the interface. Very likely, these defects propagate in the epilayer. In this case, the deuterium solubility is determined by the concentration of defects in the epilayer and at the interface and not by the boron concentration. For this reason, there is a competition between the formation of deuterium-related shallow donors and trapping of deuterium on the structural defects and this competition impedes the shallow donor formation. The exact origin of the structural defects which are detrimental for the p-type to n-type conversion is not established but it is theoretically known that dislocations act as deep hydrogen traps (binding energy=1.9-4 eV) [6]. It has also been experimentally shown that vacancy-hydrogen complexes are stable up to 1500°C [31]. Trapping of hydrogen on the defects present at the epilayer/substrate interface is very stable since annealing the sample at 1050°C, 24 h does not modify neither the deuterium distribution nor the deuterium concentration at this interface.

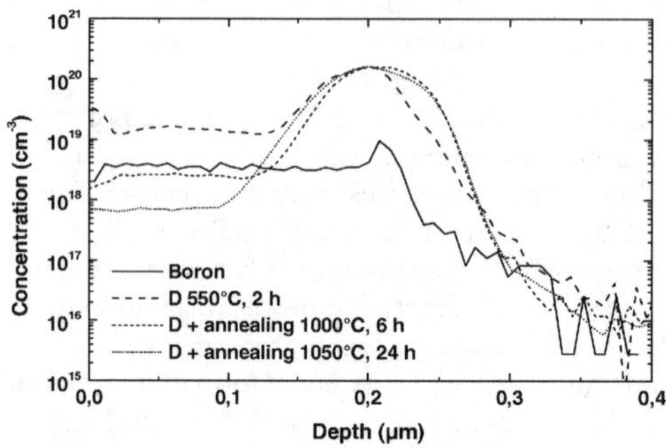

Figure 5. SIMS profiles of boron and deuterium after deuteration at 550°C, 2 h of a boron-doped epilayer containing defects at the epilayer/substrate as revealed by the strong deuterium accumulation at this interface. Annealings at 1000°C, 6 h and then 1050°C, 24 h do not significantly change this deuterium accumulation.

Evidence of deuterium diffusion in type Ib diamond

Coming back to the deuterated boron-doped epilayer of figure 1, we see that thermal annealing at 750°C, 1 h results in an out-diffusion of deuterium (figure 6). This is expected because our previous results have shown that deuterium starts out-diffusing at 720°C [26]. We also observe a clear deuterium diffusion profile in the type Ib diamond substrate containing 2×10^{19} cm^{-3} optically active isolated nitrogen atoms as deduced from infrared transmission experiments [32]. This profile can be fitted by an erfc function and the corresponding effective diffusion coefficient is 1.4×10^{-14} cm^2/s at 750°C. This is the first experimental evidence of hydrogen diffusion in type Ib diamond. Since the hydrogen acceptor level (E_c-2 eV) is below the nitrogen donor level (E_c-1.7 eV), we expect that H$^-$ is the diffusing species in type Ib diamond. For comparison, the effective diffusion coefficient of protons in an epilayer containing 2×10^{19} cm^{-3} boron is $D_{eff}(750°C)=4 \times 10^{-12}$ cm^2/s according to equation (1) [18]. This means that H$^-$ diffuses more slowly in nitrogen-doped diamond than H$^+$ in boron-doped diamond.

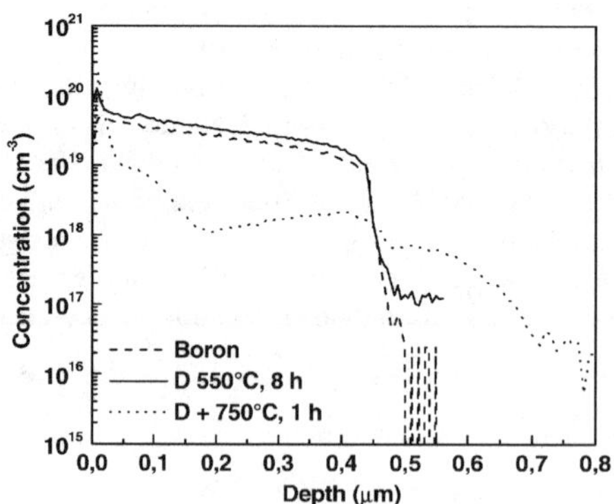

Figure 6. SIMS profiles of boron and deuterium of the sample of figure 1 after a thermal annealing at 750°C, 1 h. Under this annealing, we clearly observe a deuterium diffusion in the type Ib diamond substrate.

Previously, we attempted to detect deuterium diffusion in type Ib diamond after immersing such a sample in a deuterium plasma at 1000°C, 52 h but we failed [21]. The explanation of these differences might be found in the nature of the deuterium diffusion source. In the deuterium plasma exposed samples, we have an infinite source but the diamond is etched during the plasma exposure. We have found etching rates in the range 200-600 Å/h for p-type boron-doped diamond for diffusion temperatures in the range 500-950°C but with no clear correlation between the etching rate and the sample temperature. Qualitatively, we can understand that if the etching rate is of the order of or faster than the deuterium diffusion rate, the deuterium diffusion profiles will not be detectable.

In the case presented in figure 6, the deuterium diffusion is performed after a pre-loading deuterium step at 550°C in the boron-doped epilayer in order to create a reservoir of (B,D) complexes. Then, thermal annealing at 750°C dissociates the (B,D) complexes and out-diffuses most of the deuterium but a fraction of the deuterium diffuses deeper in the substrate. Because the dose of deuterium is not constant during this annealing, the above effective diffusion coefficient at 750°C is a lower limit.

In order to evaluate the effect of the etching rate on the deuterium diffusion profile, we have calculated the deuterium distribution C(x,t) in the presence of etching.

We start with the modified Fick's equation:

$$D\frac{\partial^2 C}{\partial t^2} = \frac{\partial C}{\partial t} - v\frac{\partial C}{\partial y} \tag{2}$$

with y=x-vt. D is the effective deuterium diffusion coefficient and v is the etching rate.

The limit conditions are: $C(y=0,t)=C_0$, $C(y,t=0)=0$ and $C(y\rightarrow\infty,t)=0$.

The solution of equation (2) is :

$$C(x,t)=\frac{C_0}{2}\left[\left(\exp\left(-\frac{v(x-vt)}{D}\right)\right)\left(erfc\left(\frac{x}{2\sqrt{Dt}}-v\sqrt{\frac{t}{D}}\right)\right)+\left(erfc\left(\frac{x}{2\sqrt{Dt}}\right)\right)\right]$$

For v=0, we find the complementary function :

$$C(x,t) = C_0\,erfc\left(\frac{x}{2\sqrt{Dt}}\right)$$

Figure 7 presents the results of our calculations for t=1 h and t=52 h with no etching and with a etching rate of 600 Å/h taking $D=1.4\times10^{-14}$ cm^2/s. This figure shows that, due to the plasma etching, the deuterium diffusion profile after a 52 h plasma exposure remains shallow and this may well explain the non observation of deuterium diffusion in type Ib diamond in the case of a deuterium plasma exposure.

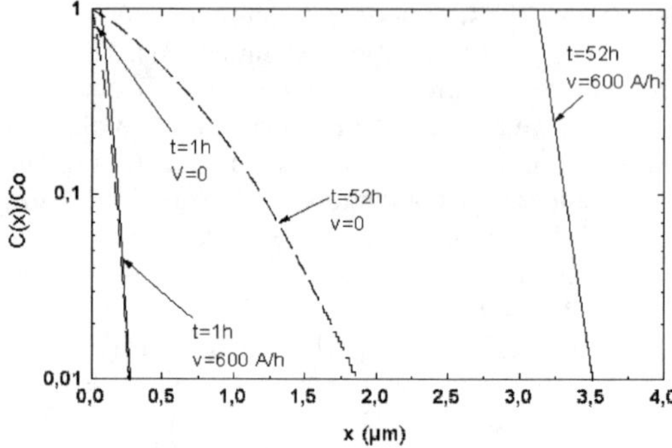

Figure 7. Calculated deuterium diffusion profiles at t=1 h and t=52 h with v=0 and v=600 Å/h taking a deuterium diffusion coefficient of 1.4×10^{-14} cm^2/s. The shift of the profile 52 h, 600 Å/h corresponds to the etched thickness.

CONCLUSIONS

We have shown that deuterium-related shallow donors can be created in deuterated boron-doped diamond giving high n-type electrical conductivity at 300 K. The ionization energy of these donors is 0.34 eV. These donors dissociate at 520°C while (B,D) pairs are stable at this temperature. Above 550°C, (B,D) pairs start to dissociate. The n-type conductivity occurs when excess deuterium is formed. This excess deuterium appears once all boron atoms are saturated and the deuterium diffuses much faster than in diamond containing empty boron traps. It is attributed to the formation of another deuterium-related complex. Under the assumption that the shallow donors are associated with the creation of the new complexes, we conclude that their formation is the result of a two step process: passivation of all boron acceptors present in the epilayers followed by the trapping of deuterium on some species. These results are coherent with our initial model of the formation of (B,D$_2$) complexes but do not preclude the formation of other specific (D,defect) complexes. The quality of the epilayers is a key point for the success of the p-type to n-type conversion because structural defects present in the epilayer will generate also a large excess deuterium attributed to the trapping of deuterium by these structural defects. These defects will trap deuterium at the expense of the deuterium-related shallow donor formation.

In addition, we have demonstrated that deuterium diffusion occurs in type Ib diamond. This observation has been possible by pre-loading deuterium in the boron-doped epilayer at 550°C in order to create a reservoir of (B,D) complexes followed by a thermal annealing at a higher temperature where (B,D) complexes dissociate and deuterium diffuses into the type Ib diamond. The effective diffusion coefficient of deuterium in type Ib diamond is equal to or above 1.4×10^{-14} cm^2/s at 750°C. The non observation of deuterium diffusion by a deuterium plasma exposure is explained by a etching of the epilayer simultaneous to the deuterium diffusion. Further experiments are needed in order to get more accurate diffusion coefficients of hydrogen in type Ib diamond. They should be performed in conditions where the dose of deuterium is kept constant during the thermal annealing step.

REFERENCES

1. J.E.Butler and D.G.Goodwin, *Properties, Growth and Applications of Diamond,* eds M.H.Nazaré and A.J.Neves, EMIS Datareviews Series n°**26**, 262 (2001).
2. P.K.Baumann and R.J.Nemanich, *J.Appl.Phys.* **83**, 2072 (1998).
3. J.B.Cui, J.Ristein, L.Ley, *Phys.Rev.Lett.* **81**, 429 (1998).
4. J.Ristein, F.Maier, M.Riedel, M.Stammer, L.Ley, *Diam.Rel.Mat.* **10**, 416 (2001).
5. K.Tsugawa, K.Kitatani, H.Noda, A.Hokazono, M.Tajima, H.Kawarada, *Diam.Rel.Mat.* **8**, 927 (1999).
6. J.P.Goss, R.Jones, M.I.Heggie, C.P.Ewels, P.R.Briddon and S.Öberg, *Phys.Rev.B* **65**, 115207 (2002).
7. L.G.Wang and A.Zunger, *Phys.Rev.B* **66**, 161202(R) (2002).
8. E.Holzschuh, W.Kündig, P.F.Meier, B.D.Patterson, J.P.F.Sellschop, M.C.Stemmet and H.Appel, *Phys.Rev.A* **25**, 1272 (1982).
9. B.Holm, K.Bonde Nielsen and B.Bech Nielsen, *Phys.Rev.Lett.* **66**, 2360 (1991).
10. S.K.Estreicher, *Mater.Science and Engin.* **R14, 7-8**, 319 (1995).
11. J.Chevallier, B.Theys, A.Lusson, C.Grattepain, A.Deneuville and E.Gheeraert, *Phys.Rev.B* **58**, 7966 (1998).
12. R.Zeisel, C.E.Nebel and M.Stutzmann, *Diam.Rel.Mat.* **9**, 413 (2000).
13. C.Saguy, C.Cytermann, B.Fizgeer, V.Richter, Y.Avigal, N.Moriya, R.Kalish, B.Mathieu and A.Deneuville, *Diam.Rel.Mat.* **12**, 623 (2003).
14. S.P.Mehandru and A.B.Anderson, *J.Mat.Res.* **9**, 383 (1994).
15. J.P.Goss, R.Jones, M.I.Heggie, C.P.Ewels, P.R.Briddon and S.Öberg, *Phys.Stat.Sol. (a)* **186**, 263 (2001).
16. Ch.Langpape, S.Fabian, Ch.Klatt and S.kalbitzer, *Appl.Phys.A* **64**, 207 (1997).
17. J.Chevallier, in *Thin Film Diamond*, eds C.Nebel and J.Ristein, Semiconductors and Semimetals, vol.77 (Academic Press, San Diego, 2004) (in press).
18. Z.Teukam, D.Ballutaud, F.Jomard, J.Chevallier, M.Bernard, A.Deneuville, *Diam.Rel.Mat.* **12**, 647 (2003).
19. C.P.Herrero, M.Stutzmann, A.Breitschwerdt and P.V.Santos, *Phys.Rev.B* **41**, 1054 (1990).
20. C.Herring and N.M.Johnson, in *Hydrogen in Semiconductors*, Semiconductors and Semimetals, vol.34 (Academic Press, San Diego, 1991) p.225.
21. J.Chevallier, F.Jomard, Z.Teukam, S.Koizumi, H.Kanda, Y.Sato, A.Deneuville, M.Bernard, *Diam.Rel.Mat.* **11**, 1566 (2002).
22. S.Koizumi, T.Teraji and H.Kanda, *Diam.Rel.Mat.* **9**, 935 (2000).
23. E.Gheeraert, S.Koizumi, T.Teraji, H.Kanda, M.Nesladek, *Diam.Rel.Mat.* **9**, 948 (2000).
24. Z.Teukam, J.Chevallier, C.Saguy, R.Kalish, D.Ballutaud, M.Barbé, F.Jomard, A.Tromson-Carli, C.Cytermann, J.E.Butler, M.Bernard, C.Baron and A.Deneuville, *Nature Materials* **2**, 482 (2003).
25. J.S.Blakemore in *Solid State Physics,* 2nd edition, Cambridge University Press 1985, p.320.
26. D.Ballutaud, F.Jomard, J.Le Duigou, B.Theys, J.Chevallier, A.Deneuville and F.Pruvost, *Diam.Rel.Mat.* **9**, 1171 (2000).

27. J.Chevallier, A.Lusson, D.Ballutaud, B.Theys, J.Chevallier, A.Deneuville, M.Bernard, E.Gheeraert and E.Bustarret, *Diam.Rel.Mat.* **10**, 399 (2001).
28. C.Saguy, R.Kalish, C.Cytermann, Z.Teukam, J.Chevallier, F.Jomard, A.Tromson-Carli, J.E.Butler, C.Baron and A.Deneuville, *Diam.Rel.Mat.* (2004) in press.
29. J.P.Goss, P.R.Briddon, S.J.Sque and R.Jones, *Phys.Rev.B* (2004) in press.
30. Y.Dai, D.Dai, D.Liu, S.Han and B.Huang, *Appl.Phys.Lett.* **84**, 1895 (2004).
31. X.Zhou, G.D.Watkins, K.M.McNamara Rutledge, R.P.Messmer and S.Chawla, *Phys.Rev.B* **54**, 7881 (1996).
32. I.Kiflawi, A.E.Mayer, P.M.Spear, J.A.van Wyk and G.S.Woods, *Phil.Mag.B* **69**, 1141 (1994).

Hydrogen diffusion in polycrystalline boron doped and undoped diamond.

D. Ballutaud[1], A. Boutry-Forveille[1], J.-M. Laroche[1], N. Simon[2], H. Girard[2], M. Herlem[2], and A. Etcheberry[2]
[1]CNRS-LPSC, 1 place Aristide Briand, 92195 Meudon cedex, France.
[2]Institut Lavoisier, Université de Versailles-St-Quentin en Yvelines, 78035 Versailles cedex, France.

ABSTRACT

Hydrogen (deuterium used as tracer) diffusion experiments were performed on undoped and boron doped diamond films ($[B] = 10^{19}$ and 10^{20} cm^{-3}) grown by plasma CVD or hot-filament assisted CVD. The samples were exposed either to a radiofrequency plasma or a microwave plasma at different temperatures between 400°C and 900°C. The deuterium profiles were analysed by secondary ion mass spectrometry (SIMS). The deuterium diffusion was explained mainly in term of trapping on intergranular defects. The passivation of boron acceptors, by B-D complex formation in the deuterium diffused superficial layers of the diamond films, was followed by electrochemical and mercury probe capacitance measurements. The results suggest a strong decrease of the free carrier density, which is in accordance with passivation of free carriers by deuterium trapping on dopant.

INTRODUCTION

Hydrogen is known to play an important role in the diamond film growth from a CH_4/H_2 precursor mixture. The diamond polycrystalline films are highly defective, and present a high density of grain boundaries or dislocations, and incorporated hydrogen in semiconductors is known to terminate dangling bonds and passivate both shallow and deep levels. Furthermore, the hydrogen present at the diamond surface induces a superficial highly p-type conductive layer, the origin of which is not yet completely understood [1].

In a previous paper [2], it has been shown that the hydrogen concentration in an as-grown CVD polycrystalline diamond film with 1 μm grain size is about 10^{19} cm^{-3}. This paper deals with new data on the diffusion of hydrogen (deuterium used as tracer) in polycrystalline doped and undoped diamond films. The deuterium diffusion profiles are analysed, and capacitive measurements [3] are used to evidence the passivation of boron acceptors by boron-deuterium complex formation.

EXPERIMENTAL

Boron ($[B] = 2\times10^{19}$ and 2×10^{20} cm^{-3}) doped polycrystalline diamond films (respectively 0.3 and 0.2μm grain size) were deposited on monocrystalline silicon substrates in a hot-filament-assisted chemical vapour deposition reactor supplied with diborane and methane in hydrogen and provided by CSEM [4]. Some undoped polycrystalline films grown by plasma CVD using a 0.5 % CH_4/H_2 mixture (950 W, 6000 Pa, 780° C) on a monocrystalline silicon substrate were also used throughout this study (1μm crystal size) [5]. The hydrogen concentrations in the original as-grown layers are respectively10^{20} cm^{-3} in the boron doped layers and 2×10^{19} cm^{-3} in the undoped layers [2].

The samples were submitted to hydrogen (deuterium used as tracer) remote radio-frequency (r.f.) plasma (power density between electrodes 0.4 W.cm^{-2}) at 500°C, or to deuterium microwave plasma at 600° (400 W, 2500 Pa) and 800°C (500 W, 6000 Pa) during 5 hours.

The deuterium diffusion profiles were analysed with a CAMECA IMS 4f Secondary ion mass spectrometer with a Cs$^+$ primary ion beam. Quantifications of deuterium concentration were achieved by using undoped CVD diamond standards implanted with a known dose of deuterium.

Just before the capacitance-voltage measurement experiments, the boron doped diamond samples were boiled in a (HNO_3; $HClO_4$; H_2SO_4) mixture to take off the high electrical conductivity hydrogen terminated diamond surface [1]. The capacitance–voltage C(V) measurements were carried out in 2M H_2SO_4 solutions using a Fabelle-CNRS potentiostat 1.107 MHz, and a lock-in amplifier (EG&G 5208) in a classical three electrode set-up. Capacitance voltage measurements were also performed using a mercury probe coupled to a capacitance Keithley bridge at 1 MHz.

RESULTS AND DISCUSSION

<u>Hydrogen diffusion from a plasma</u>

The deuterium diffusion profiles obtained from a r.f. plasma (500°C, 5h) are shown in figure 1, for the undoped sample (curve a), the 2×10^{19} cm^{-3} boron doped sample (curve b), and the 2×10^{20} cm^{-3} boron doped sample (curve c). The deuterium depth profile presents an enhanced deuterium concentration in the surface region. This subsurface accumulation may be due to deuterium trapping on plasma induced defects or free deuterium diffusion.

Generally, the presence of traps for hydrogen reduces the apparent diffusivity, leading to an effective coefficient [6]:

$$D_{eff} = D_H \left([H]/[H_T] \right)$$

where $[H]/[H_T]$ is the fraction of free hydrogen. In this case, it may be observed experimentally an exponential fall of the depth profile:

$$[H] = [H_0] \exp(-\alpha x)$$

x being the diffusion depth and with the argument α of that exponential varying with the trapping concentration [6][7]:

$$\alpha = \sqrt{4\pi R [n_T]}$$

where $[n_T]$ is the total initial trap concentration and R is the capture radius of this trap for hydrogen. Then d, the mean free path between trapping events is given by $1/\alpha$ [7].

The experimental profiles in figure 1 follow an exponential decay at lower concentrations. The dotted lines represent a least square fit with a characteristic length $d=1/\alpha$. If it is supposed that the capture radius for traps is the same for the three samples, it may be concluded that the mean free path d between traps decreases with the grain size, i.e. decreases when the concentration of defects n_T at grain boudaries increases. From the shape of the deuterium concentration profiles, it appears that the deuterium diffusion is governed mainly by trapping on defects due to the polycrystalline nature of the diamond layers. The deuterium trapping on boron dopant, which leads to the formation of a boron deuterium complex at this diffusion temperature (500°C) [8], does not affect the deuterium diffusion profile shape and is completely masked by deuterium trapping onto deep traps.

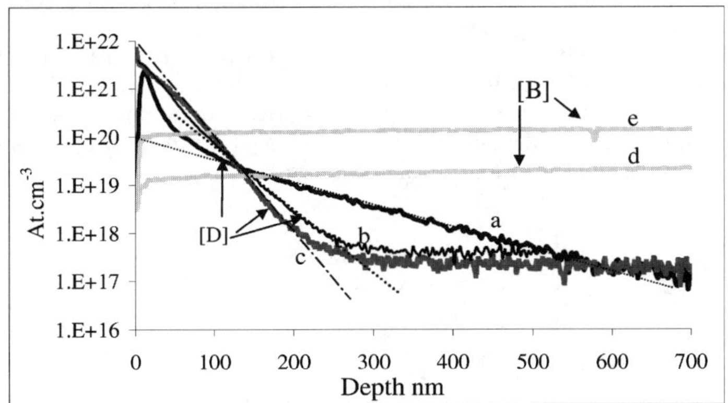

Figure 1. Deuterium diffusion profiles obtained at 500°C (r.f. plasma) in undoped polycrystalline diamond (curve a), in 10^{19} cm^{-3} and 10^{20} cm^{-3} boron doped polycrystalline diamond (curves b and c); the boron level for samples b and c has been reported on the figure (curves d and e).

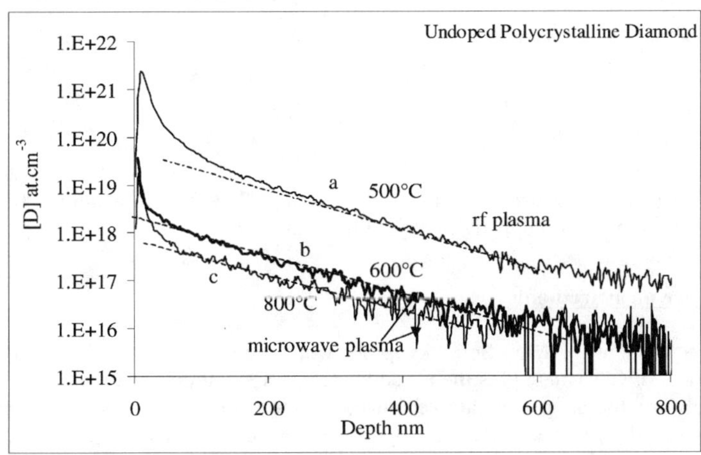

Figure 2. Deuterium diffusion profiles obtained at 500°C (r.f. plasma)(curve a), 600°C and 800°C (microwave plasma) (curves b and c) in undoped polycrystalline diamond.

Figure 2 represents the results of deuterium diffusion experiments performed on undoped polycrystalline diamond respectively at 500°C (r.f. plasma) (curve a), 600°C (curve b) and 800°C (curve c) (microwave plasma) for 5 hours. The use of microwave plasma instead of rf plasma causes a decrease of the deuterium surface concentration. The dotted lines on curves a, b and c represent the fits to an exponential decay, and result in the same mean free path value independant of temperature, which confirms that the low deuterium concentration region represents deuterium trapping onto deep traps.

The deuterium diffusion profiles performed at 600°C and 800°C (microwave plasma, 5 hours) (curve a and b) for the 10^{19} cm^{-3} boron doped sample are presented in figure 3. As it is observed above for the undoped sample, it appears from the shape of the deuterium concentration profiles that the deuterium diffusion is governed mainly by trapping on defects due to the polycrystalline nature of the diamond layers.

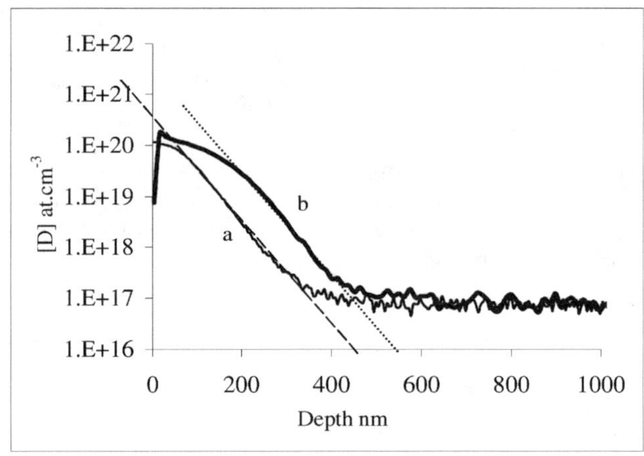

Figure 3. Deuterium diffusion profiles obtained at 600 °C (curve a) and 800°C (curve b) in 10^{19} cm^{-3} boron doped polycrystalline diamond.

Capacitance measurements

Electrochemical Mott-Schottky plots were recorded for the 10^{19} cm^{-3} boron doped sample. Plots of C^{-2}, where C is the capacitance, vs. applied polarization, are given in figure 4, respectively for the as-grown sample (curve a), the same sample after deuteration at 500°C (curve b) and the same sample after deuteration at 800°C (curve c). The as-grown sample reveals a classical Mott-Schottky behavior (curve a), which allows us to calculate a free carrier concentration of 3.8×10^{19} cm^{-3}. After deuteration at 500°C (curve b), the representation of C^{-2} is completely flat, revealing a nearly insulating behavior of the upper layers of the material.

This is in accordance with previous results obtained in diamond, where it has been shown that the passivation effect of free carriers occurs by hydrogen trapping on boron dopant [10]. After deuteration at 800°C, the dopant is partially passivated, and a free carrier density of 2.7×10^{19} cm^{-3} is calculated from the C^{-2} plot, which confirms that the boron-hydrogen complex is stable up to 750°C [8].

Plots of C^{-2}, using the mercury probe, are given in figure 5, for the 10^{19} cm^{-3} boron doped sample, before (curve a) and after deuteration at 800°C (curve b). From the Mott-Schottky plot, the free carrier concentration computes to 3×10^{19} cm^{-3} before deuteration and to 2×10^{19} cm^{-3} after deuteration. Similarly to the electrochemical C(V) measurements, it evidences a partial passivation of dopant by deuterium.

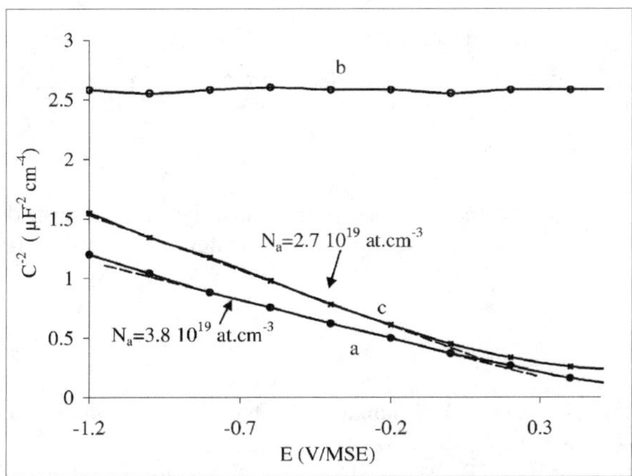

Figure 4. Electrochemical Mott-Schottky plots for a 10^{19} cm^{-3} boron doped polycrystalline diamond layer; curve a: as-grown sample.; curve b: deuterated at 500°C (r.f. plasma); curve c: deuterated at 800°C (microwave plasma).

CONCLUSION

From the shape of the deuterium concentration profiles it appears that deuterium diffusion is governed mainly by trapping on defects in polycrystalline diamond layers. The deuterium mean free path between traps decreases with the grain size, i.e. decreases when the bulk concentration of defects due to grain boudaries increases.

Although the deuterium trapping on boron dopant, which leads to a boron deuterium complex, does not affect the deuterium diffusion profile and is completely masked by deuterium trapping on deep traps, the dopant passivation by deuterium has been evidenced by electrochemical and mercury probe capacitance measurements. Results show a complete passivation of the dopant in the deuterated superficial layers of the sample when the deuteration is performed at 500°C, and a partial passivation when deuteration is performed at 800°C.

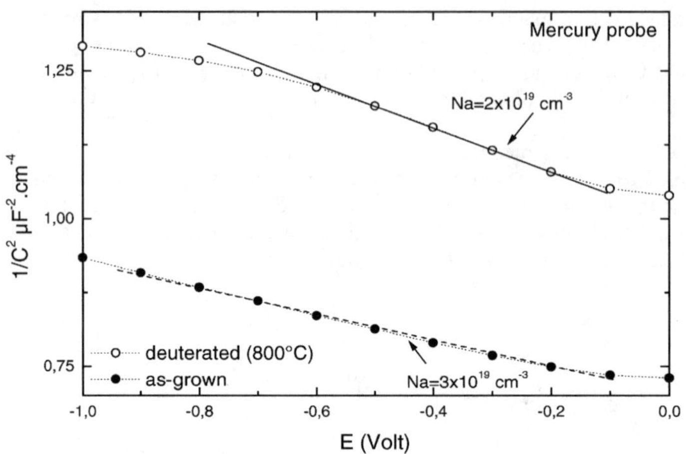

Figure 5. Plots of C^{-2} using the mercury probe for a 10^{19} cm^{-3} boron doped polycrystalline diamond layer, before (curve a) and after deuteration at 800°C (curve b)

REFERENCES

1. S. Yamanaka, D. Takeuchi, H. Watanabe, H. Okushi and K. Kajimura, Appl. Surf. Science **159-160**, 567 (2000).
2. D. Ballutaud, F. Jomard, B. Theys, C. Mer, D. Tromson and P. Bergonzo, Diam. Rel. Mater. **10**, 405 (2001)
3. D. A. Tryk, K. Tsunozaki, Tata N. Rao and A. Fujishima, Diam. Rel. Mater. **10**, 1804 (2001).
4. CSEM, Jaquet-Droz 1, CH-2007 Neuchâtel (Switzerland).
5. C. Jany, F. Foulon, P. Bergonzo, A. Tardieu and A. Gicquel, Proceeding of the fifth international symposium on diamond materials (Paris, September 1997), Electrochem. Soc. Proc. **97-32**, 208 (1998).
6. Hydrogen in crystalline semiconductors, S. J. Pearton, J. W. Corbett and M. Stavola, (Springer Series in Material Science, Spinger Verlag, Berlin, 1992) pp. 200-233.
7. N. H. Nickel, W. B. Jackson and J. Walker, Phys. Rev. B **53 (12)**, 7750 (1996).
8. J. Chevallier, A. Lusson, D. Ballutaud, B. Theys, F. Jomard, A. Deneuville, M. Bernard, E. Gheeraert and E. Bustarret, Diam. Rel; Mater. **10**, 399 (2001).

Non-metastable recombination induced reactions involving hydrogen in SiC.

Yaroslav Koshka, Bharat Krishnan, Michael S. Mazzola.
*Department of Electrical & Computer Engineering, Mississippi State University, Box 9571,
Mississippi State, MS 39762, USA*

ABSTRACT

A brief survey of some of the recent results of recombination-induced defect reactions
involving hydrogen in 4H and 6H-SiC is given. A variety of outcomes of such reactions have
been observed in hydrogenated 4H and 6H-SiC polytypes under optical excitation at reduced
temperature. A few different non-metastable hydrogen-defect complexes can form, including
hydrogen complexes with Al and B acceptors, hydrogen complex with Si vacancy, as well as
some other non-identified complexes. Electrical measurements indicated strong recombination-
induced passivation of the electrical activity of aluminum and boron acceptors in SiC. This
passivation resulted in the reduction of the net free hole concentration and even inversion of the
conductivity type. It is suggested that optical excitation causes a long-range migration of
hydrogen followed by its capture in one or another kind of defect complexes. Further insight in
the formation of specific complexes as a result of recombination-induced defect reactions is
provided by thermal admittance spectroscopy. Energy gap levels in the regions that remained p-
type after hydrogenation as well as in the regions where the conductivity was inverted by the
recombination-induced passivation are investigated.

INTRODUCTION

Defect migration and defect reactions enhanced or induced by carrier capture and
recombination fascinated semiconductor researchers for many years (see for example Ref.1-3).
Associated phenomena are particularly complex and diverse in wide band-gap semiconductors
[4]. Electronic mechanism of enhanced defect migration has also been reported for hydrogen.
Such phenomena were observed for example in hydrogenated amorphous silicon (α-Si:H) [5].

The first evidence of recombination-enhanced defect reactions (REDRs) in SiC was
obtained by Dean and Choyke for a hydrogen complex with Si vacancies (V_{Si}-H) [6]. Excitation
of SiC samples with above band-gap light at low temperature resulted in metastable quenching of
the V_{Si}-H photoluminescence (PL). The corresponding REDR was attributed to a transition
between a stable and a metastable configuration of the hydrogen-defect complex [6,7]. Non-
metastable REDRs in SiC not related to hydrogen were also reported. Recombination-induced
generation and growth of stacking faults was found in forward-biased SiC pn diodes [8].

Our recent results indicated that electronic mechanism can produce athermal migration of
hydrogen in p-type 4H and 6H-SiC polytypes. Athermal migration of hydrogen at low
temperatures (15 K) resulted in the formation of a variety of non-metastable complexes with
defects and impurities [9-11]. The process named recombination-induced passivation (RIP)
exhibited complex temperature behavior, with the efficiency of the new complex formation
getting significantly reduced when approaching room temperature. In many cases, the process
resulted in significant changes of the net free carrier concentration and conductivity type.

In this paper, a brief overview of recent results of recombination-induced migration of
hydrogen in SiC is presented. Different experimental observations are analyzed to establish the

likely mechanism responsible for this process. Results of electrical characterization are analyzed in order to establish the degree of acceptor passivation caused by the RIP process as well as the depth of the passivated layer. The preliminary results of the thermal admittance spectroscopy (TAS) in SiC samples with conductivity inverted by the RIP process are presented.

EXPERIMENTAL DETAILS

4H and 6H-SiC epitaxial layers of 2-10 μm thickness were grown by atmospheric pressure chemical vapor deposition (CVD) [12]. Aluminum and boron dopants were introduced during the growth using trimethylaluminum and a solid boron-nitride source 12] respectively. The net doping concentration in different p-type samples was from $5x10^{15}$ cm^{-3} to above $3x10^{17}$ cm^{-3}. The epilayers were subjected to hydrogen plasma in an inductively coupled plasma (ICP) etching system at a source power of 750 W, an rf power of 100 W, and a pressure of 50 mTorr.

A special attention was paid to the design of the plasma-hydrogenation process. Different designs of the wafer carrier were used to provide different hydrogenation temperatures. The temperature was estimated to be in the range from below 100^0C (for efficient sample cooling) to a few hundreds of degrees (for intentionally bad thermal contact between the sample and the wafer carrier).

The PL measurements were conducted using an Ion Ar+ laser line of 301 nm. The total excitation power in all the experiments did not exceed 25 mW. PL was measured at around 15 K. Optical excitation for RIP following plasma hydrogenation was conducted at the same conditions that were used for measuring PL. Planar Ti Schottky diodes were fabricated for CV and thermal admittance spectroscopy measurements.

RESULTS AND DISCUSSION

Different outcomes of the RIP process and the proposed explanation are schematically illustrated in figure 1. Hydrogen trapped in the material after plasma hydrogenation but not yet involved in the formation of any stable complexes (so-called "free" hydrogen) can migrate athermally under above bandgap excitation. During this migration, hydrogen experiences defect reaction to form (1) complexes with Al acceptors (Al-H), (2) complexes with B acceptors (B-H), (3) complex responsible for $4B_0$ luminescence, (4) complex with a Si vacancy (V_{Si}H), as well as some other complexes that are yet to be identified. The efficiency of the plasma-hydrogenation and the features of the RIP process following plasma hydrogenation strongly depend on the hydrogenation conditions. It was established that the design of the wafer holder of the ICP system as well as the thermal contact of the sample to the holder drastically influenced the temperature of the process, which in turn produced different hydrogenation outcomes. Almost no acceptor passivation effect could be observed after 2 hours of low-temperature plasma hydrogenation (Fig. 2(a)). Low degree of hydrogen incorporation was also confirmed by PL measurements. The earlier reported hydrogen-related features in the PL spectrum [11,13] (i.e., reduced intensity of Al bound exciton (Al-BE) emission, appearance of the boron-related $4B_0$ PL and the lines due to the hydrogen complex with Si vacancies - V_{Si}-H) were very weak after low-temperature hydrogenation (Fig. 3(a)). In contrast, hydrogenation at conditions when the sample was significantly heated by hydrogen plasma resulted in formation of a passivated layer with concentrations of electrically active aluminum (or boron) acceptors significantly reduced near the surface (Fig. 2(b)). The depth and degree of passivation strongly depended on acceptor

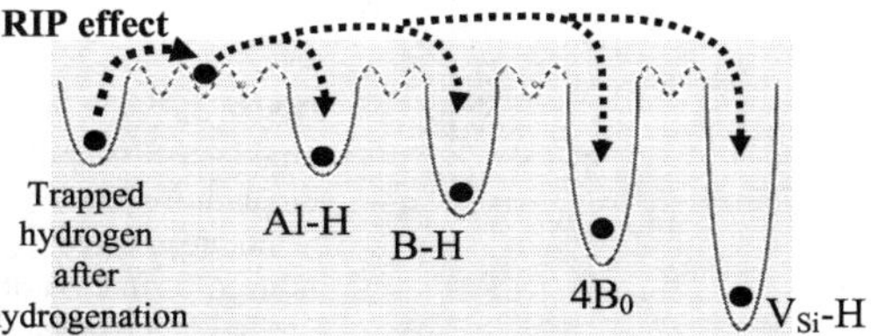

Figure 1. Schematic illustration of the proposed mechanism of the RIP process. Following hydrogenation, the above bandgap excitation at low temperatures causes release of hydrogen from the trapping sites near the surface, its athermal migration and trapping by more stable sites (e.g., Al and B acceptors, silicon vacancies, etc.)

concentration, which was previously shown to be the result of trap limited diffusion of hydrogen [14]. The effect of this high-temperature hydrogenation on the PL spectrum (not shown) was pronounced, including (1) a few times decrease of the intensity of Al-BE confirming strong passivation of Al acceptors observed in figure 2(b), (2) appearance of strong hydrogen-complex-related $4B_0$ line, and (3) appearance of intense V_{Si}-H lines that become the dominating features in the PL spectra of strongly hydrogenated samples. However, the seeming absence of any hydrogen incorporation by low-temperature plasma processing turned out to be deceptive. The samples hydrogenated at low temperatures (but at the same density of the hydrogen plasma) demonstrated the strongest changes under optical excitation at reduced temperatures (i.e., RIP effect). Fast reduction of Al-BE photoluminescence and growth of hydrogen related lines ($4B_0$ and V_{Si}-H lines) were induced in these samples by excitation with above bandgap light (Fig. 3(b) and (c)). After the samples had been removed from the cryostat and the doping profile was measured in locations subjected to optical excitation (RIP), a strong passivation of acceptors (Al or B depending on the type of the sample) similar to that in figure 2 was observed [11]. A sufficiently prolonged RIP processing regularly resulted in complete inversion of the conductivity type near the surface [11].

The samples hydrogenated at higher temperatures also exhibited recombination-induced formation of hydrogen-defect complexes. However, the relative magnitude of this effect was much smaller than after low-temperature hydrogenation because a significant amount of hydrogen-defect complexes had already been formed during plasma-hydrogenation, leaving less opportunity for the recombination-induced processes.

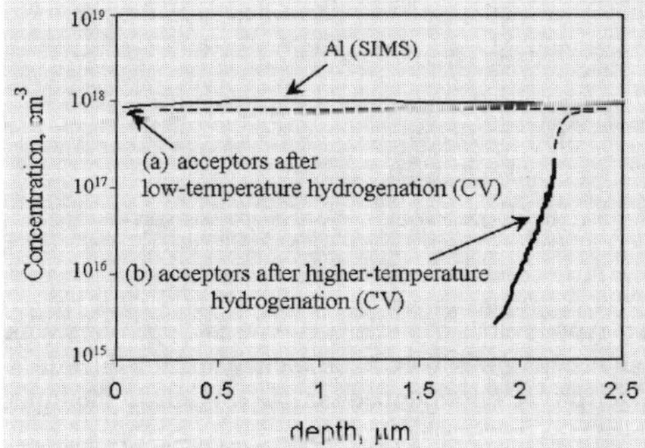

Figure 2. Changes in the concentration of electrically active Al acceptors after 2 hours of plasma-hydrogenation when the sample was efficiently cooled during the processing **(a)** and when the sample was allowed to hit up to a few hundreds of degrees **(b)**. No electrical passivation took place during hydrogenation at low temperature.

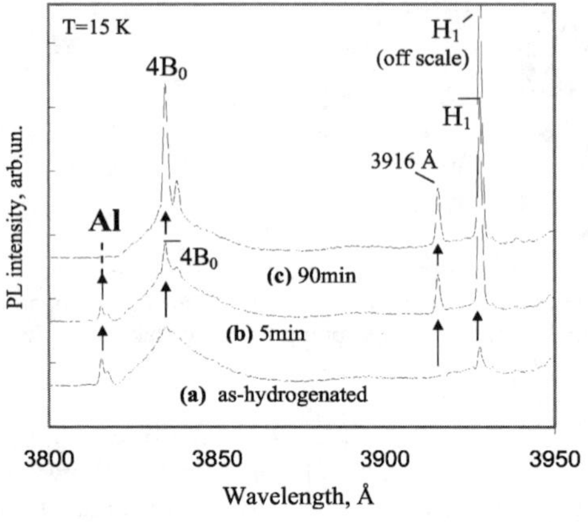

Figure 3. Weak changes in PL of Al-doped 4H-SiC epilayer after low-temperature plasma hydrogenation **(a)** and pronounced effect of recombination-induced formation of hydrogen-related luminescence centers **(b)** and **(c)**. Reduction of the Al line and growth of 4B$_0$ and H$_1$ PL under the above band-gap light is shown after 5 min **(b)** and 90 min of excitation **(c)**.

Similar experiments with deuterium replacing hydrogen (to enhance sensitivity of SIMS technique) are yet to be conducted. The hydrogen concentration after plasma hydrogenation and RIP was normally close or lower than the detection limit of SIMS technique. For that reason, the Capacitance-Voltage (CV) measurements were used as the primary technique to investigate the depth distribution of the acceptor passivation efficiency and the thickness of the inverted region. Analysis of the depth profile of the passivated layer after different RIP times or different intensities of RIP excitation [11] suggests that the passivation proceeds from the surface of the wafer progressively deeper in the epilayer. This result is consistent with our suggestion of recombination-induced migration of hydrogen from the surface followed by hydrogen trapping by acceptors and other defect centers.

CV doping profiles measured in the regions of the samples inverted by the RIP process are shown in figure 4. The thickness of the samples in figures 4 (a) and (b) were around 2 μm and 3 μm respectively. The shape of the depth profile in figure 4(a) was typical for initially strongly Al and B doped p-type samples. Shapes of the depth profiles similar to that shown in figure 4(b) were typically observed in lighter doped p-type samples. Ideally, after the majority of the

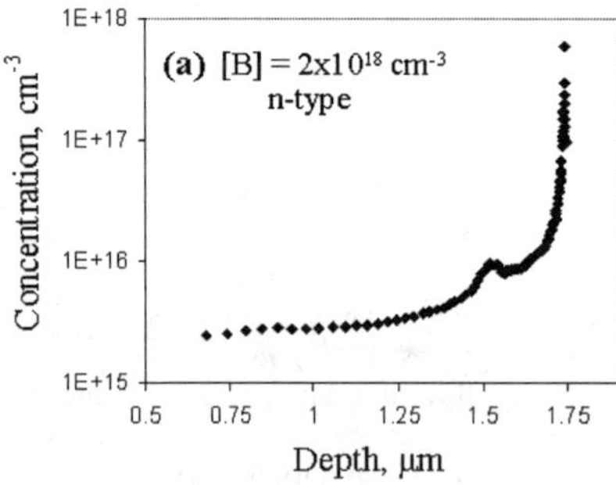

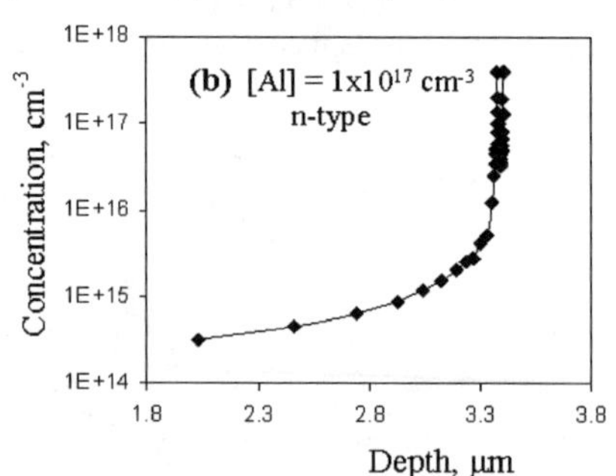

Figure 4. CV depth profiles measured in n-type regions of initially p-type SiC epilayers with conductivity inverted by RIP process. Both acceptor passivation efficiency and the thickness of the inverted region are higher in the stronger p-type boron doped sample **(a)** than in the lighter p-type Al doped sample **(b)**.

acceptors are passivated with hydrogen by the RIP process, the CV will measure the net donor concentration (N_D-N_A). Using two-dimensional simulation of the CV response of the investigated structure, it was established that both profiles do not give true impurity distribution. Careful analysis allowed us to suggest that the profile in figure 4(a) is likely to correspond to the true net donor concentration closer to the surface. The apparent growth of the measured concentration with depth is most likely a measurement artifact. The best fit of these data was obtained with the assumption of the epilayer being inverted through its entire 2 μm thickness.

The profile in figure 4(b) is a result of the inverted layer being (1) thinner than the thickness of the epilayer and (2) fully depleted. The space-charge region penetrates to another depletion layer at the junction between inverted (n-type) and passivated but non-inverted (p-type) regions. In this case, the change of the capacitance with the reverse bias has nothing to do with incremental penetration of the depletion region deeper in the material. The best fit of the CV curve and the apparent doping profile allowed us to suggest that the most likely structure responsible for the profile in figure 4(b) is a 2 μm weakly inverted n-type near-surface region (with N_D-N_A of the order of 1×10^{14} cm^{-3} followed by p-type region with net acceptor concentration gradually increasing away from the junction.

The stronger inversion in p-type samples that have had higher acceptor concentration before hydrogenation and RIP (figure 4(a)) can have two explanations. First, hydrogen incorporation and trapping in this type of material is expected to be higher [14]. This is true for thermal diffusion, which in our experiment is responsible for initial hydrogen distribution during plasma treatment. The same trend may also hold for recombination-induced athermal migration.

Second, the concentration of compensating nitrogen donors in strongly boron doped p-type materials used in our study was also higher as determined by SIMS. Nearly complete passivation of dominating acceptors would leave more donors to contribute to n-type conductivity.

Currently, we only have evidences that RIP effect on conductivity comes from passivation of Al and B acceptors. There is yet no evidence that the hydrogenation and/or RIP create new donors. However, this possibility can not be completely excluded. Thermal admittance spectroscopy measurements in the inverted regions are attempted in order to answer this question. In addition, it offers a unique possibility to probe the shallow defect levels in the top portion of the bandgap of the samples that were p-type before RIP.

The TAS spectrum of a typical Al-doped p-type epilayer of figure 4(b) after plasma hydrogenation but prior to RIP is shown in figure 5. The p-type conductivity of this sample is

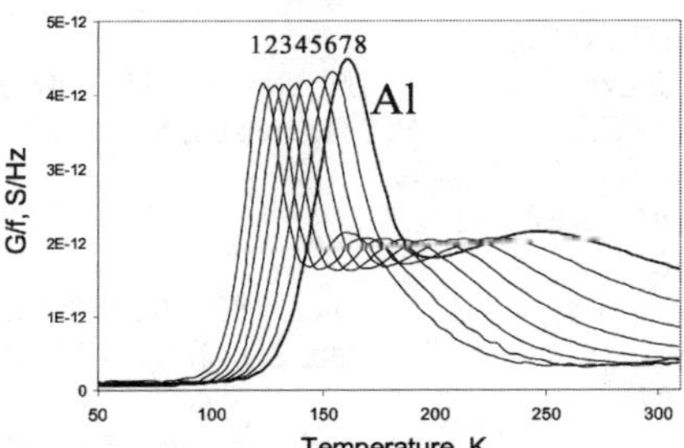

Figure 5. Admittance spectra for hydrogenated Al-doped p-type 4H-SiC epilayer. The measurement frequency changes from (1) 6 kHz to (8) 800 kHz.

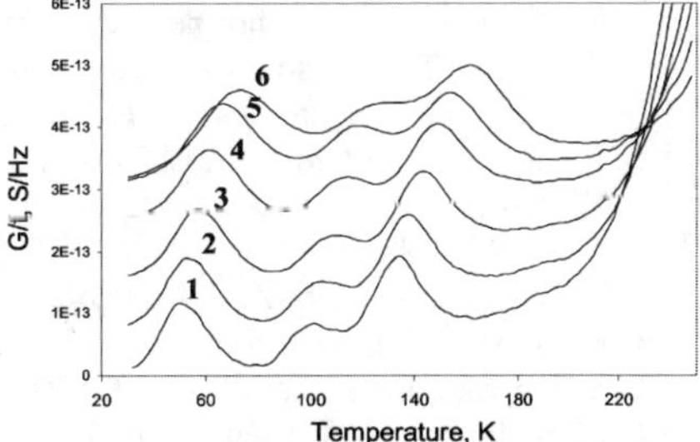

Figure 6. TAS measured in the sample with conductivity inverted by OIH. The measurement frequency changes from (1) 25 kHz to (6) 1 MHz.

determined by the aluminum acceptors with E_A of 0.20 or 0.22 eV depending on the model for temperature dependence of the impurity capture cross section used in calculations [15]. The plasma-hydrogenation conditions were such that the net free hole concentration changed very little in comparison to that prior to hydrogenation.

The TAS spectrum of the conductivity-inverted region of the sample is shown if figure 6. A few shallow levels as well as one deep level beyond 250 K (not shown) were detected. The extracted positions of the bandgap levels in the conductivity inverted sample are 0.04 eV (0.05 eV), 0.13 eV (0.15 eV), 0.21 eV (0.23 eV), and a deeper level at 0.4-0.55 eV (0.55-0.6 eV).

Presently, reliable identification of the admittance peaks in the inverted region of the sample is not available. The higher energy peaks may be the result of the proximity of the junction between inverted and non-inverted region (i.e., it could be related to acceptors in the p-region). However, the lower temperature peaks offer interesting information about levels close to the conduction band of the inverted region.

CONCLUSION

A preliminary insight in the degree of recombination-induced passivation in p-type SiC as well as in the depth of athermal migration is obtained. In addition, TAS offers opportunity to investigate shallow levels in the top portion of the valence band present in p-type epilayer before hydrogenation as well as those introduced by the RIP.

ACKNOWLEDGEMENT
This work was supported by the Office of Naval Research, Grant No. N00014-03-1-0581, Dr. C. Wood program manager.

REFERENCES
1. B. L. Gregory, J. Appl. Phys.36, 3765 (1965)
2. A.M. Stoneham, Philos. Mag, 36, 983 (1977).
3. D.V. Lang, L.C. Kimerling, and S.Y. Leung, J. Appl. Phys. 47, 3587 (1976).
4. M.A. Rizakhanov, M.M. Khamidov, Yu.N. Emirov, Crys. Inorg. Mater. 36 (12), 1200 (2000).
5. P. V. Santos, N. M. Johnson, R. A. Street, M. Hack, R. Thompson, and C. C. Tsai, Phys. Rev. B. 47 (16), 10244 (1993).
6. P.J. Dean and W.J. Choyke, Adv. Phys. 26, 1 (1977).
7. A. Henry, T. Egilsson, I.G. Ivanov and E. Janzen, Mat. Sci. For. 338-342, 651 (2000).
8. J.P. Bergman, H. Jakobsson, L.Storasta, F.H.C. Carlsson, B. Magnusson, S. Sridhara, G. Pozina, H. Lendenmann, E. Janzén, Mater. Sci. For. 389-393, 9 (2002).
9. Y. Koshka, M. S. Mazzola, Appl. Phys. Lett, 79(6), 752 (2001).
10. Y. Koshka, Appl. Phys. Lett. 82, 3260 (2003).
11. Y. Koshka, A. Los, M. S. Mazzola, I. Sankin, Physica B, 340–342, 180–183 (2003).
12. M. S. Mazzola, S. E. Saddow and A. Schöner, Mater. Sci. Forum, 264-268, 119 (1998).
13. Y. Koshka, Phys. Rev. B 69, 0352051 (2004).
14. M. S. Janson, A. Hallén, M. K. Linnarsson, and B. G. Svensson, Phys. Rev. B 64, 195202 (2001).
15. W. Kaindl, M. Lades, N. Kaminski, E. Niemann, and G. Wachutka, Jorn. of Electron. Mater. 28 (3), (1999), p. 154.

Hydrogen in Oxides

Anharmonicity in the Vibrational Modes Associated with H-H, N-H, O-H, and C-H bonds

Sukit Limpijumnong

School of Physics, Institute of Science, Suranaree University of Technology,
Nakhon Ratchasima 30000, Thailand.

ABSTRACT

The anharmonic contributions to the vibrational frequency of various H-H, N-H, O-H, and C-H bond configurations are investigated using first principles calculations. While harmonic approximation can be accurately used to calculate the local vibrational mode of most microscopic configurations, the configurations involving strong X-H bond are exceptions because of the exceptionally light mass of the H atom. We will explicitly present the magnitude of the anharmonic contributions to the total frequencies associated with X-H bonds in free molecules as well as in semiconductors defects.

INTRODUCTION

Hydrogen is one of the most abundant and well-known element in universe. In molecular form[1], it can form strong bond with various elements. It plays an important role as defects in semiconductor [2-4] and in semiconductor alloys [5]. It also plays an important role on semiconductor surfaces [6]. In semiconductors, an interstitial H atom and a H_2 molecule are widely studied. As an interstitial atom, it has a tendency to form a strong bond with the host atom or impurity. Infrared spectroscopy, which probes the vibration excitation of local vibrational modes, is a widely used technique to detect H [7-9]. The technique can be used to gain information about hydrogen-related microscopic structures of defects in semiconductors. In calculating the vibration frequency, most of the time the harmonic approximation is used[3][10-14]. Under harmonic approximations, the anharmonic part of the vibration is ignored. In a system composed of only heavy atoms, the anharmonic part of the vibration is indeed very small. Even for rather light molecules such as O_2 or NO molecules, the anharmonicity of the stretch vibration is reasonably small. For examples, including anharmonic corrections for O_2 or NO stretch vibration would result in the reduction of the vibrational frequency by only about 20 cm^{-1}

out of the total frequency of ~2000 cm^{-1}. *However, for systems containing* H *atom, which is exceptionally light, the anharmonicity of the vibration is sizable.* Applying anharmonic corrections to a calculated vibration associated with an X-H bond (in this study $X \equiv$ H, N, O, or C), results in a reduction the total frequency by 150-400 cm^{-1}, or about 10% of the total frequency. Our calculated anharmonic part of the vibration frequency for various X-H bonds in both free molecules form and in defects in semiconductor form are shown in Fig. 1. The plot shows that anharmonic contribution to the vibration frequency is inversely proportional to the bond distance and more enhanced in semiconductors. The large anharmonicity in the vibration of X-H bonds has a root from the fundamental quantum properties of the vibration. Under *a harmonic approximation*, one must assumes *small vibration amplitude* such that the potential-displacement curve can be reasonably described with a perfect parabola. This approximation is, however, unreasonable for the vibrations associated with X-H bonds. For X-H bonds, the amplitude of the zero point vibrations is already extended into the region where the potential energy curve is significantly anharmonic. Anharmonicity is, therefore, inevitably needed for accurate calculations of vibration frequencies associated with such bonds.

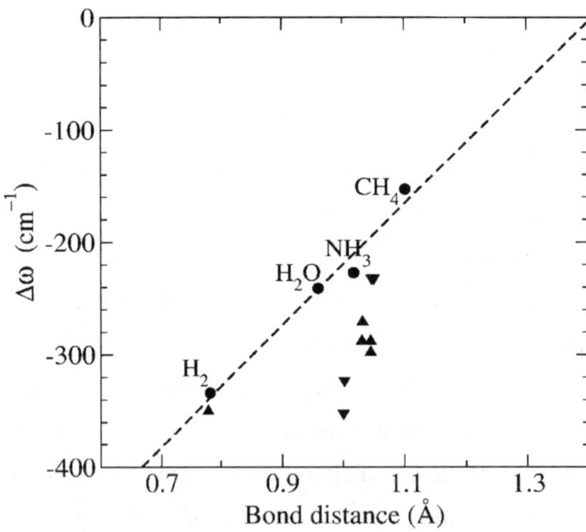

Figure 1. Calculated anharmonic part of the vibration frequency associated with various X-H bonds as a function of the bond distance. The filled circles represent the results of free molecules. The dashed line is a linear fit of the free molecule results. The up and down triangles represent anharmonic part of the vibration frequencies associated with various X-H bonds in GaN [21-22] and ZnO [23], respectively.

METHODS

First principles calculations

We employ density-functional theory (DFT) in the local density approximation (LDA) [15] and *ab initio* pseudopotentials (as implemented in the FHI98 code [16]) or ultrasoft pseudopotentials (as implemented in the VASP code [17]), with a plane-wave basis set. The H defects in semiconductors (GaN and ZnO) are calculated in the wurtzite phase of the materials (with theoretically optimized volume and structural parameters) using a supercell approach with the supercells containing 32 or 36 atoms. Free molecules are calculated using a supercell approach with a large supercell size of $10 \times 10 \times 10$ Å to ensure minimal inter-cell interactions. Brillouin-zone integrations were carried out following the Monkhorst-Pack scheme[18]. The k-points are generated by a regularly spaced mesh of $n \times n \times n$ within the first Brillouin-zone. Some points in this set are related by symmetry; those points that are not related by symmetry, are said to be located in the irreducible part of the zone. Convergence tests indicated that a $2 \times 2 \times 2$ sampling suffices for our 32- and 36-atom supercell and 1 to 2 sampling k-points suffice for the $10 \times 10 \times 10$ Å supercell of a free molecule.

Calculation of frequencies for vibrational modes

For a given system of H atom(s) in a semiconductor or in a free molecule, we start by calculating the atomic configurations and total energy of stable configurations. Our results show that in all cases, the H atom forms a strong bond with another atom (or with another H atom in the case of H_2 molecule). We will therefore focus on the vibration mode of such bond. In case of dimer molecule as well as an interstitial H atom in semiconductor, there exist only one such strong bond and the stretch vibration of the bond can be simplified to a one-dimension single-particle oscillator *via* the general reduced mass approach. In other more complicated free molecules, when we focus only on the *symmetric* stretch mode, we can simplify the problem into a one-dimension single-particle oscillator using similar idea as the general reduced mass approach, describing in the next section. We simulate the vibrational frequency by tracing the energy-displacement curve of the vibration. This has been done by calculating energies for configurations in which the H atom is displaced from its equilibrium site in both compression and extension directions of vibrations. We typically include about 10-20 displacements, with magnitude up to $\pm 30\%$ of the bond length. For the case of an oscillator in semiconductor, no host atoms are displaced from their equilibrium sites in the calculations. It is reasonable to fix host atoms in their respective equilibrium positions because H is much lighter than the host atoms. In addition, the vibration frequencies of the vibrational modes under study are much higher than the frequency of the crystal phonon. Limpijumnong and Van de Walle[19] have compared the vibration frequency obtained by freezing host atoms with the one obtained from the dynamic matrix approach and found a good agreement, providing that the reduced mass is used in the prior case. To proper taking care of the anharmonic effect, we calculated the anharmonic contributions of the frequency by fitting the calculated energy-displacement curve with a fourth-degree polynomial,

$$V(s) = \frac{k}{2}s^2 + \alpha s^3 + \beta s^4. \tag{Eq. 1}$$

We calculate the vibration frequencies as follows. The transition energy from the ground state to the first excited state can be written as $\Delta E_{0-1} = E_1 - E_0 = \hbar\omega = \hbar(\omega^0 + \Delta\omega)$, where ω is the total frequency, ω^0 is the harmonic frequency, and $\Delta\omega$ is the anharmonic contribution. E_0 and E_1 are the solutions to the one-dimensional single-particle Schrödinger equation

$$\left(-\frac{\hbar^2}{2\mu}\nabla^2 + V(s)\right)\psi(s) = E\psi(s).$$

To solve the Schrödinger equation, we can either (1) use the perturbation theory [20] to obtain an approximated analytical solution to the Schrödinger equation with above potential or (2) solve the Schrödinger equation numerically. Under perturbation theory, the frequency is

$$\omega = \omega^0 + \Delta\omega = \sqrt{\frac{k}{\mu}} - 3\frac{\hbar}{\mu}\left(\frac{5\alpha^2}{2k^2} - \frac{\beta}{k}\right). \tag{Eq. 2}$$

For the stretching mode vibration of a N-H bond, Limpijumnong, Northrup, and Van de Walle[21] have tested the quality of perturbation results against the direct numerical solutions and found a good agreement to within a few 10 cm^{-1}. Most of the values presented in this article are obtained from the perturbation calculations. However, to clearly illustrate that the large anharmonicity for the vibration of X-H bond has a root from the fundamental quantum properties of the vibration, we will present the calculation of NH$_3$ symmetric stretch vibration using the numerical method.

<u>**Reduced mass (μ)**</u>

Traditionally, the concept of reduced mass is introduced so that one can view a two-body problem as if it were a single-particle problem. The reduced mass of a two-body system is defined as the inverse of the sum of each inverse mass, i.e. $\frac{1}{\mu} = \frac{1}{m_1} + \frac{1}{m_2}$, where m_1 and m_2 are the masses of particle 1 and particle 2, respectively. The reduced mass approach can be readily applied to dimer molecules. As an oscillator, a dimer molecule can be simplified to a one-dimension single-particle oscillator with a mass μ, and a coordination variable s (which is the bond distance between the two atoms). For example, the H_2 molecule has $\mu = m_H / 2$ and the amplitude of the vibration can be described by the H-H bond distance s with an equilibrium bond distance $s_0 \approx 0.7 \, \text{Å}$. The concept of reduced mass can be applied to the X-H bond in bulk semiconductors. Under an assumption that only the pair of atoms composing the oscillator can move and other host atoms are fixed, the reduced mass of the oscillator can be defined the same way as a dimer molecule. For detailed explanation regarding H oscillator in semiconductors, please see Ref. [19][21-22].

We can extend the concept of reduced mass to simplify other molecule such as H_2O or NH_3 to a one-dimension single-particle oscillator. However, we have to take into account the geometry of the molecule as well as the mode of vibration. As an example, we will illustrate the simplification of the symmetric stretch mode vibration of an NH_3 molecule.

The geometry of an NH_3 molecule is illustrated in Fig. 2. In the symmetric stretch vibration mode, all three H atoms are vibrating simultaneously with the same amplitude along the direction of their respective N-H bond. In other words, all three N-H bonds are equally and simultaneously compressed and extended. The problem can be simplified into a one-dimension single-particle oscillator, if we can describe both kinetic and potential terms in the Schrödinger equation using only a coordination variable and a simplified reduced mass. Since all three N-H bonds are compressed and extended simultaneously, the potential energy can be described by a single variable s, which is a single N-H bond distance (as shown in Fig. 2).

The kinetic energy can be written as

$$3\left(\frac{1}{2}m_{\mathrm{H}}\dot{r}_{\mathrm{H}}^2\right)+\frac{1}{2}m_{\mathrm{N}}\dot{r}_{\mathrm{N}}^2 = 3\left(\frac{1}{2}\mu\dot{s}^2\right),\qquad\qquad\text{(Eq. 3)}$$

where m_{H} and m_{N} are the masses of H and N atoms, r_{H} and r_{N} are the amplitude of the

displacement of H and N atoms, s = N-H bond distance, and μ is the simplified reduced mass of

the oscillator that we need to find. Note that the kinetic energy per N-H bond is one-third of the

total kinetic energy, i.e. $\frac{1}{2}\mu\dot{s}^2$. The vertical component of the r_{H} is $r_{\mathrm{H}}\cos\theta$. In order to fulfill

the requirement that the center of mass of the molecule remain fixed, the displacement of the N

atom (r_{N}) needs to equal to $3\left(\frac{m_{\mathrm{H}}}{m_{\mathrm{N}}}\right)r_{\mathrm{H}}\cos\theta$. Therefore, the N-H bond distance changes by

(ignoring the small change in the angle θ)

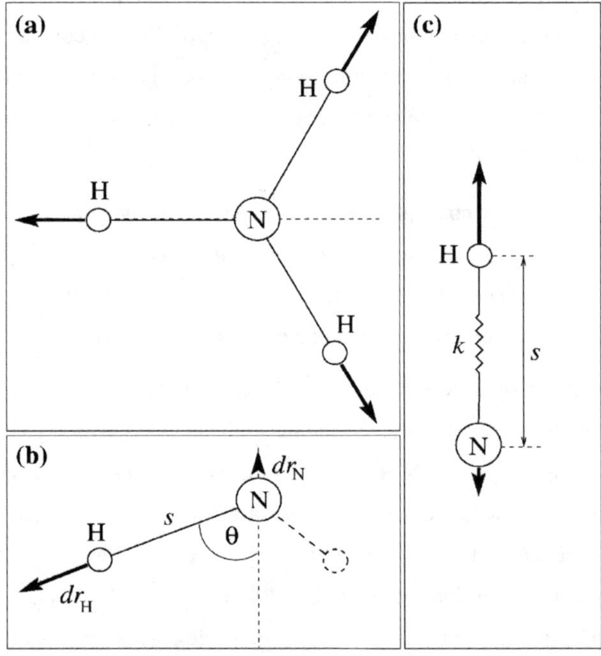

Figure 2. Schematic representation of the symmetric stretch mode of NH$_3$. (a) Top view; (b) Side view, cutting through one of the N-H bonds. (c) shows a simplified oscillator.

$$(ds)^2 = (r_\mathrm{H} \sin\theta)^2 + (r_\mathrm{H} \cos\theta + r_\mathrm{N})^2$$

$$= (r_\mathrm{H} \sin\theta)^2 + \left(r_\mathrm{H} \cos\theta + 3\frac{m_\mathrm{H}}{m_\mathrm{N}} r_\mathrm{H} \cos\theta \right)^2$$

$$= \left[1 + \cos^2\theta \left(\frac{6m_\mathrm{H}}{m_\mathrm{N}} + \frac{9m_\mathrm{H}^2}{m_\mathrm{N}^2} \right) \right] r_\mathrm{H}^2.$$

Substitute ds and r_N in terms of r_H and $m_\mathrm{N} = 14 m_\mathrm{H}$ into (Eq. 3) and derive for the reduced mass μ, we get

$$\mu = \frac{1 + 0.2143 \cos^2\theta}{1 + 0.4745 \cos^2\theta} m_\mathrm{H}. \qquad\qquad (\text{Eq. 4})$$

We can check the validity of (Eq. 4) using two asymptotic conditions. (1) If the angle θ were equal to $90°$, the N atom would remain immobile in the symmetric stretch mode, corresponding to an infinitely heavy mass for the nitrogen atom and a reduced mass $\mu = m_\mathrm{H}$. Substituting $\theta = 90°$ into (Eq. 4), we indeed get $\mu = m_\mathrm{H}$. (2) If the angle θ were equal to $0°$, the system is equivalent to a dimer molecule with a N atom on one end and an atom with the mass of $3m_\mathrm{H}$ on the other end. The reduced mass of the system would be $2.471 m_\mathrm{H}$ or $\frac{2.471}{3} m_\mathrm{H} = 0.824 m_\mathrm{H}$ per oscillator. Substituting $\theta = 0°$ into (Eq. 4), we indeed get $\mu = 0.824 m_\mathrm{H}$. For an actual NH_3 molecule we substitute $\theta = 69°$ and get the reduced mass $\mu = 0.969 m_\mathrm{H}$. In this way, the symmetric stretch vibration mode of a NH_3 molecule is simplified into a one-dimension single-particle oscillator with the coordination variable s and a simplified reduced mass μ.

RESULTS

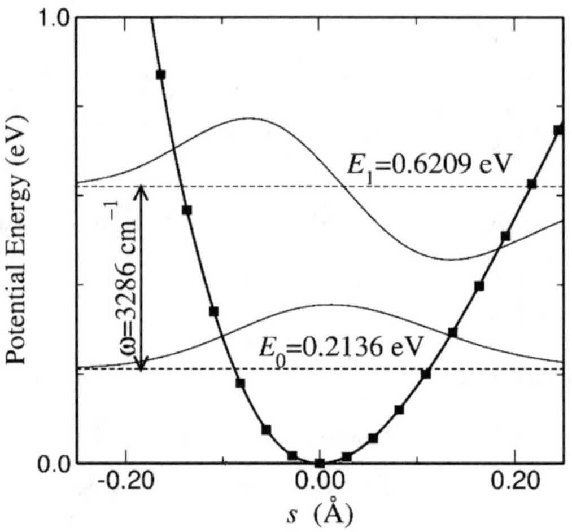

Figure 3. Calculated potential energy curve for a simplified N-H oscillator in the symmetric bond stretch mode of NH_3. The solid squares are the calculated results and the thick solid line is the fourth-order polynomial fit. The ground state and the first excited state energies are shown as thin-dashed lines, and the corresponding wave functions are shown in thin solid lines.

The calculated potential curve for the stretch mode vibration (per N-H bond) of a free NH_3 molecule, which represents the typical potential curve of other oscillators under study here, is shown as a thick solid line in Fig. 3. It is clear that the potential curve is largely deviated from a perfect parabola. The fitting parameters for the potential energy curve is shown in Table I. We can solve the Schrödinger equation with this potential numerically [21] (using the simplified reduced mass of NH_3 $\mu = 0.969m_H$). The resulting ground state and first excited state wave functions are shown as thin solid lines and the associated energies are shown as thin dashed lines. It is clear that the wave functions have appreciative values far out to the amplitude as large as 0.3 Å, the region where the anharmonicity is evident. This is a clear sign that anharmonic effect has large influence on the vibration frequency of these systems. Therefore, one cannot simply calculate such vibration frequencies under harmonic approximation using small vibration amplitude and hope to get reliable results. Since the typical anharmonic contribution of such vibrations is around 200 cm^{-1}, it is reasonable to think that any prior harmonic prediction of better agreement with the observed frequency is fortuitous.

Substituting the fitting parameters $k/2$, α, and β obtained from the NH_3 potential energy curve into (Eq 2), we find $\Delta\omega = -227$ cm^{-1} and $\omega = 3286$ cm^{-1}.

Calculations of the vibration of other free molecules and H defects in semiconductors show the magnitude of the anharmonic contributions to vary between 150-400 cm^{-1} depending strongly on bond type and detailed configuration (Table I). $\Delta\omega$ is largest in the case of H-H bond with the value between 330-390 cm^{-1}. For the N-H and O-H bonds, the anharmonic contributions are slightly smaller and varied in a wider range of 200-350 cm^{-1}. For C-H bond, we only studied the CH_4 free molecule and found a relatively smaller $\Delta\omega$ of -153 cm^{-1}. The magnitude of the anharmonic contribution roughly goes with the inverse bond length (See Fig. 1). The linear fits of the free molecule cases corresponds to

$$\Delta\omega = -761 + 541d ,$$

where d is the oscillator bond distance in Å and $\Delta\omega$ is in cm^{-1}. In general, the anharmonicity of X-H bonds is enhanced in semiconductors. However, there are some cases that specific microscopic configuration leads to a smaller anharmonic contribution. For example, the H^+ at the bond center (BC) site in GaN [21].

With the proper inclusion of the anharmonic part of the vibration, the total calculated vibration frequencies are in satisfactory agreement with the observed values to within 100-150 cm^{-1} (except for H-H bonds where the calculations underestimated the observed values by about 300 cm^{-1}).

Calculations using two slightly different computation details, i.e. the pseudopotential as implemented in FHI98 code *versus* the ultrasoft pseudopotential as implemented in VASP code, give reasonable agreement of the anharmonic contributions (see Table I).

Table I Calculated vibration properties of selected free molecules and H defects in wurtzite GaN and ZnO. $k/2$, α, and β are the second-, third-, and fourth-order coefficients in the polynomial fit of the potential energy curve (Eq. 1), in eV/Å^n, where n is the order of the coefficient. ω^0, $\Delta\omega$ and ω are the harmonic component, anharmonic component and total vibration frequency, respectively (see text). All frequencies are in cm^{-1}. For free molecules, only the results of the FHI98 code[a] are used for the plot (Fig. 1).

Configuration	Bond type	d (Å)	$k/2$	α	β	ω^0	$\Delta\omega$	ω	ω (exp)
Free molecules									
H_2	H-H	0.782	16.31	−34.10	34.88	4194	−334	3860[a]	4161[c]
		0.765	17.00	−38.91	45.30	4281	−390	3891[b]	
H_2O	O-H	0.979	26.21	−61.03	57.19	3848	−241	3606[a]	3657[d]
		0.975	25.77	−64.28	74.53	3815	−257	3559[b]	
NH_3	N-H	1.018	22.29	−54.56	58.02	3507	−227	3286[a]	3337[d]
		1.018	20.73	−45.95	48.71	3389	−195	3194[b]	
CH_4	C-H	1.102	16.05	−32.66	34.25	2941	−153	2789[a]	2917[d]
H defects in wurtzite GaN (Ref. [21][22])									
H_2	H-H	0.779	15.42	−32.91	34.07	4081	−350	3731[a]	
H^+ at AB$_{N\perp}$ site	N-H	1.045	17.40	−48.48	57.24	3170	−298	2872[a]	
H^+ at AB$_{N\parallel}$ site	N-H	1.045	16.72	−47.74	60.18	3108	−288	2820[a]	
Mg-H at AB$_{N\perp}$ site	N-H	1.031	19.24	−51.11	56.49	3333	−288	3045[a]	3125[e]
Mg-H at OA$_{\parallel}$ site	N-H	1.032	19.31	−50.51	55.32	3339	−271	3068[a]	
H defects in wurtzite ZnO									
H^+ at AB$_{O\perp}$ site	O-H	1.003	19.93	−54.21	63.57	3379	−323	3056[b]	
H^+ at AB$_{O\parallel}$ site	O-H	1.001	19.83	−56.47	70.19	3370	−352	3018[b]	
N-H at AB$_{N\perp}$ site	N-H	1.050	17.11	−40.19	44.13	3144	−232	2912[b]	
N-H at AB$_{N\parallel}$ site	N-H	1.047	17.43	−41.33	47.26	3173	−232	2941[b]	

[a] Calculate using the pseudopotential (as implemented in the FHI98 code)
[b] Calculate using the ultrasoft pseudopotential (as implemented in the VASP code)
[c] Ref. [24]
[d] Ref. [25]
[e] Ref. [8],[26]

CONCLUSION

We have shown that the vibration of H-H, N-H, O-H, and C-H bonds contains a large degree of anharmonicity (typically about 10% of the overall frequency). This anharmonicity has a root from the fundamental quantum properties of the vibration mainly due to the exceptionally light mass of the H atom. We have calculated the anharmonic contributions for various H-H, N-H, O-H, and C-H oscillators in forms of free molecules as well as semiconductor defects. The anharmonic contributions of all systems under study have a negative sign with the magnitude inversely proportional to the bond distance and more enhanced in semiconductors.

ACKNOWLEDGMENTS

This work was supported by the Thai Research Fund under Contract No. BRG4680003. I thank C.G. Van de Walle, J.E. Northrup, S.B. Zhang, S.-H. Wei, and M.D. McCluskey for fruitful discussions and suggestions.

REFERENCES

[1] E. McCartney, *Absorption and emission by atmospheric gases: The physical processes* (John Wiley & Sons, 1983).

[2] C.G. Van de Walle and J. Neugebauer, Nature **423**, 626 (2003).

[3] J. Neugebauer and C.G. Van de Walle, Phys. Rev. Lett. **75**, 4452 (1995).

[4] C. G. Van de Walle, Phys. Rev. Lett. **85**, 1012 (2000); S. Limpijumnong, J. E. Northrup, and C. G. Van de Walle, Phys. Rev. Lett. **87**, 205505 (2001).

[5] A. Janotti, S. B. Zhang, and S.-H. Wei, Phys. Rev. Lett. **88**, 125506 (2002); A. Janotti, S. B. Zhang, S.-H. Wei, and C. G. Van de Walle, Phys. Rev. Lett. **89**, 086403 (2002).

[6] C. G. Van de Walle and J. Neugebauer, Phys. Rev. Lett. **88**, 066103 (2002).

[7] M.D. McCluskey, S.J. Jokela, K.K. Zhuravlev, P.J. Simpson, and K.G. Lynn, Appl. Phys. Lett. **81**, 3807 (2002).

[8] B. Clerjaud, D. Côte, A. Lebkiri, C. Naud, J. M. Baranowski, K. Pakula, D. Wasik, and T. Suski, Phys. Rev. B 61, 8238 (2000).

[9] N.H. Nickel and K. Fleischer, Phys. Rev. Lett. **90**, 197402 (2003).

[10] A. F. Wright, Phys. Rev. B **60**, 5101 (1999).

[11] A. Bosin, V. Fiorentini, and D. Vanderbilt, in Gallium Nitride and Related Materials, edited by F.A. Ponce *et al.*, Mater. Res. Soc. Symp. Proc. No. **395** (Materials Research Society, Warrendate, PA, 1996), p. 503.

[12] Y. Okamoto, M. Saito, and A. Oshiyama, Jpn. J. Appl. Phys. **35**, L807 (1996).

[13] V.J.B. Torres, S. Öberg, and R. Jones, MRS Internet J. Nitride Semicond. Res. **2**, 35 (1997).

[14] C. J. Fall, R. Jones, P. R. Briddon, and S. Öberg, Mater. Sci. Eng., B **82**, 88 (2001).

[15] P. Hohenberg and W. Kohn, Phys. Rev. B **136**, B864 (1964); W. Kohn and L. J. Sham, *ibid.* **140**, A1133 (1965).

[16] M. Bockstedte, A. Kley, J. Neugebauer, and M. Scheffler, Comput. Phys. Commun. **107**, 187 (1997).

[17] G. Kresse and J. Furthmüller, Comput. Mater. Sci. **6**, 15 (1996).

[18] H.J. Monkhorst and J.D. Pack, Phys. Rev. B **13**, 5188 (1976).

[19] S. Limpijumnong and C.G. Van de Walle, in *GaN and Related Alloys*, edited by U. Mishra *et al.*, MRS Symposia Proceedings No. 639 (Materials Research Society, Pittsburgh, 2001), p. G4.3.

[20] L.D. Landau and E.M. Lifshitz, *Quantum Mechanics* 3rd ed. (Pergamon, Oxford, 1977), p. 136.

[21] S. Limpijumnong, J.E. Northrup and C.G. Van de Walle, Phys. Rev. B **68**, 075206 (2003).

[22] S. Limpijumnong and C.G. Van de Walle, Phys. Rev. B **68**, 235203 (2003).

[23] S. Limpijumnong *et al.* (unpublished).

[24] B.P. Stoicheff, Can. J. Phys. **35**, 730 (1957).

[25] T. Shimanouchi, Tables of Molecular Vibrational Frequencies, Consolidated Volume 1, NSRDS NBS-39.

[26] W. Götz, N.M. Johnson, D.P. Bour, M.D. McCluskey, and E.E. Haller, Appl. Phys. Lett. **69**, 3725 (1996).

Infrared Spectroscopy of Hydrogen in ZnO

M.D. McCluskey and S.J. Jokela
Department of Physics, Washington State University,
Pullman, WA 99164-2814, U.S.A.

ABSTRACT

Zinc oxide (ZnO) has shown great promise as a wide band gap semiconductor with optical, electronic, and mechanical applications. Recent first-principles calculations and experimental studies have shown that hydrogen acts as a shallow donor in ZnO, in contrast to hydrogen's usual role as a passivating impurity. The structures of such hydrogen complexes, however, have not been determined. To address this question, we performed vibrational spectroscopy on bulk, single-crystal ZnO samples annealed in hydrogen (H_2) or deuterium (D_2) gas. Using infrared (IR) spectroscopy, we have observed O-H and O-D stretch modes at 3326.3 cm^{-1} and 2470.3 cm^{-1} respectively, at a sample temperature of 14 K. These frequencies are in good agreement with the theoretical predictions for hydrogen and deuterium in an antibonding configuration, although the bond-centered configuration cannot be ruled out. The IR-active hydrogen complexes are unstable, however, with a dissocation barrier on the order of 1 eV.

INTRODUCTION

Zinc oxide (ZnO) is a wide-bandgap semiconductor that has attracted tremendous interest as a blue light emitting material, a buffer layer for GaN-based devices [1], and a transparent conductor [2]. The stability of excitons in ZnO results in a very high quantum efficiency at temperatures of 300 K and higher, making it an ideal active material for the emission of blue to UV light in high-temperature environments. Theoretical work has predicted ferromagnetism above room temperature for Mn-doped ZnO, an important requirement for spintronic devices [3]. In addition to these device applications, ZnO has a potential application as a scintillator detector, due to the high cross section for nuclear reactions between ^{64}Zn and fast neutrons [4].

As-grown ZnO is nearly always *n* type. Recent theoretical work has demonstrated that hydrogen is a shallow donor in ZnO [5], raising the possibility that hydrogen donors may be introduced into the bulk during growth or processing [6,7]. Experimental results on muonium implanted into ZnO [8] and electron-nuclear resonance measurements on lightly doped, *n*-type ZnO [9] have provided evidence that hydrogen is indeed a shallow donor. In order to determine the microscopic structure of hydrogen donors, in previous work, we used infrared (IR) spectroscopy to measure the local vibrational modes (LVMs) arising from these complexes [10]. By comparing with *ab initio* calculations [5], it was proposed that the O-H complex has an antibonding orientation. However, the bond-centered orientation could not be ruled out. Two possible models for the O-H complex are shown in Fig. 1.

We have also performed polarized IR spectroscopy at room temperature [11]. These measurements indicate that the dipole of the O-H complex lies at an angle of approximately 112° to the *c* axis of wurtzite ZnO. No dipoles were observed that were oriented parallel to the *c* axis. Recent work by Nickel *et al.* [12] and Lavrov *et al.* [13] have reported hydrogen-related LVMs that are different from what is reported here.

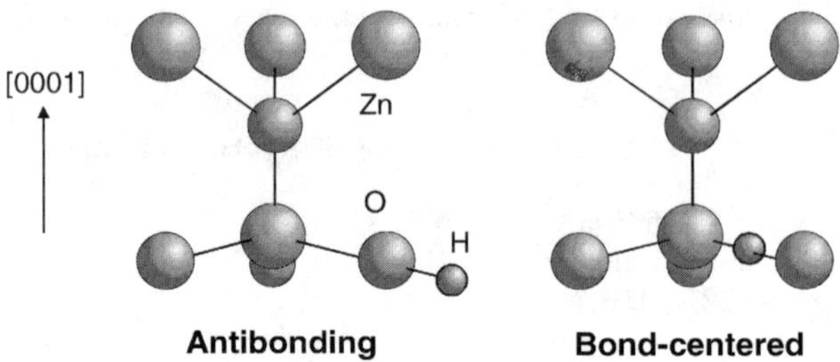

Figure 1. Two possible models for hydrogen donors in ZnO. Theoretically, the O-H complexes may also be aligned along the c axis.

EXPERIMENTAL DETAILS

Bulk, single crystal, c-cut samples of ZnO were obtained from Cermet, Inc [14]. As received, these samples have a free electron concentration of $n \sim 2 \times 10^{17}$ cm^{-3}. Hydrogen diffusion experiments were performed by sealing a quartz ampoule, evacuated and backfilled with 2/3 atm of hydrogen gas. Along with the sample, the ampoule also contained ZnO powder, in order to minimize damage to the sample surface due to the reduction of ZnO by hydrogen. The ampoule was placed in a horizontal furnace and annealed at a temperature of 720°C for a duration of 10 hr. For the last 10 min of the annealing schedule, the temperature was raised to 850°C. The ampoule was then quenched to room temperature by dropping it into water. The sample was retrieved by breaking the ampoule. After hydrogen diffusion, the free electron concentration increased to 5.5×10^{17} cm^{-3}, consistent with the claim that hydrogen in ZnO is a shallow donor.

Room-temperature Hall-effect measurements (MMR Technologies, Inc.) were performed in the van der Pauw geometry, using a 1 Tesla electromagnet. IR spectroscopy was performed using a Bomem DA8 Fourier transform IR (FTIR) spectrometer and an InSb detector. The O-H LVM was measured at room temperature and low temperatures (~14 K). For the low temperature measurements, we used a Janis STVP liquid-helium continuous-flow cryostat with wedged ZnSe windows.

DISCUSSION

Local vibrational modes

Using infrared IR spectroscopy, we have observed O-H and O-D stretch modes at 3326.3 cm^{-1} and 2470.3 cm^{-1} respectively, at a sample temperature of 14 K (Fig. 2). These modes are in agreement with those reported in Ref. 10. We did not observe the O-H modes reported by Lavrov *et al.* [13]. An unusual feature is that the deuterium peak is approximately 10 times weaker than the hydrogen peak, although the annealing conditions were the same. One possible reason for this may be the presence of hydrogen in the as-received samples. Another explanation may be that there is a greater diffusion barrier for deuterium than for hydrogen.

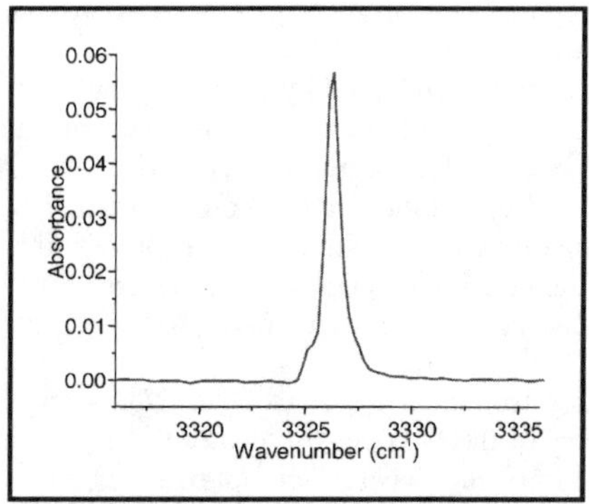 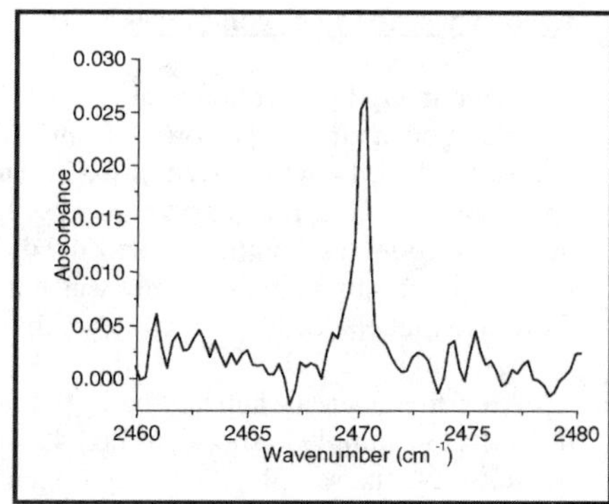

Figure 2. IR absorption spectra of O-H (left) and O-D (right) complexes in ZnO, at a temperature of 14 K.

Instability of O-H complexes

At room temperature, the area of the O-H peak gradually decreases, with a time constant of approximately one month. This behavior indicates that the O-H complexes that we observe are unstable. To investigate the kinetics in more detail, measurements were performed on a sample that had been annealed in a H_2/D_2 gas mixture. This sample contained O-H and O-D peaks. The intensities of these peaks were measured over the course of 50 days. The results are plotted in Fig. 3. The decays are exponential, with time constants of 550 and 814 hr for O-H and O-D complexes, respectively. The slower decay of the O-D complex is consistent with the lower attempt frequency of deuterium as compared to hydrogen.

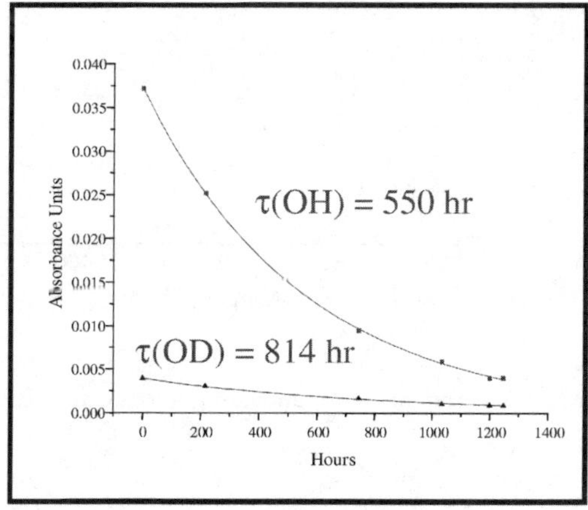

Figure 3. Area of O-H and O-D peaks as a function of time.

Recovery of the O-H complexes

After an O-H absorption peak vanishes, the peak can be recovered by annealing the ZnO sample in open air. To perform this anneal, the ZnO was placed in a quartz tube that was open at one end. The tube was moved into a horizontal furnace that had been pre-heated to 710°C. After approximately 20 s, the quartz tube was removed and dropped into water in order to rapidly quench to room temperature. The ZnO did not come into contact with the water. It should be noted that the sample temperature was not well defined during this annealing treatment. A thermocouple placed in the vicinity of the sample measured the air temperature to be approximately 520°C.

After this open-air anneal, the O-H peak increased from nearly zero to 1/2 to 2/3 its original height. An example is shown in Fig. 4. The recovery of the O-H peak indicates that hydrogen did not leave the sample; rather, hydrogen had gone into an "invisible" complex that is more stable than the O-H complex.

Hall-effect measurements on the sample before the open-air anneal showed a free electron concentration of $n = 2.0 \times 10^{17}$ cm^{-3}, which is similar to the concentration in as-received samples. After annealing in open air, the concentration increased to $n = 3.0 \times 10^{17}$ cm^{-3}. These results indicate that the invisible hydrogen complex is electrically neutral. However, more systematic measurements will be required to make that statement definitive. It has been proposed that the invisible hydrogen is in the form of H_2 molecules [15], a hypothesis that is consistent with our observations.

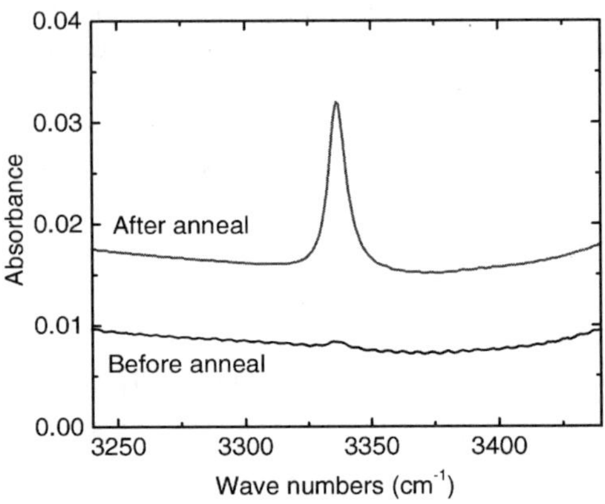

Figure 4. O-H LVM peak before and after annealing in open air. These IR spectra were taken at room temperature.

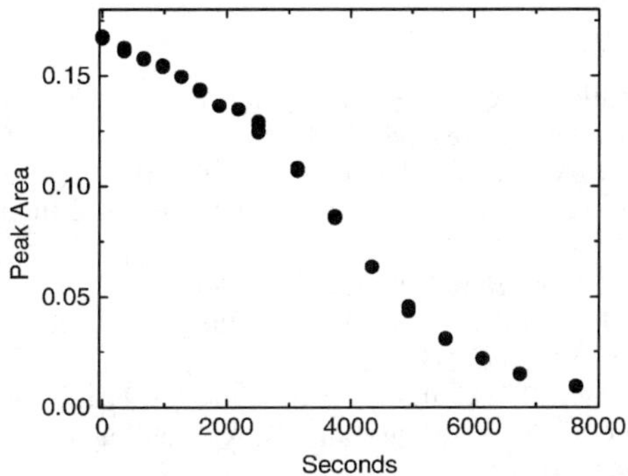

Figure 5. Plot of O-H peak area (measured at room temperature) as a function of annealing time. The annealing temperature was 100°C.

<u>Decay of O-H peak at 100°C</u>

To investigate the kinetics of the O-H decay in more detail, samples were annealed at a temperature of 100°C by boiling in water. Although there exists the possibility of introducing hydrogen from H_2O, the diffusion and solubility is low such that the O-H peak should not be affected by that process.

As shown in Fig. 5, the decay of the O-H complex at 100°C is non-exponential. The reasons for this unusual behavior are not immediately obvious. One possible explanation is that, for early times, there is a source of hidden hydrogen that forms O-H complexes. Eventually, however, the O-H peak decays to zero, as was observed for room-temperature annealing.

CONCLUSIONS

ZnO annealed in hydrogen contains O-H complexes. The O-H and O-D bond-stretching modes were measured by IR spectroscopy. At room temperature, however, these complexes decay over the course of several weeks. The complexes can then be resurrected by rapidly annealing in open air. Preliminary electrical measurements indicate that the "invisible" hydrogen is electrically neutral whereas the IR-active O-H complex is a shallow donor.

ACKNOWLEDGMENTS

The authors would like to thank M. Stavola and C.G. Van de Walle for helpful discussions. This work was supported by NSF Grant No. DMR-0203832.

REFERENCES

1. R.J. Molnar, in <u>Semiconductors and Semimetals</u> **57** (Academic Press, New York, 1999), p. 1.
2. T. Minami, *MRS Bulletin* **25** (8), 38 (2000).
3. T. Dietl, H. Ohno, F. Matsukura, J. Cibert, and D. Ferrand, *Science* **287**, 1019 (2000).
4. P.J. Simpson, R. Tjossem, A.W. Hunt, K.G. Lynn, and V. Munné, *Nucl. Instrum. Meth. A* **505**, 82 (2003).
5. C.G. Van de Walle, *Phys. Rev. Lett.* **85**, 1012 (2000).
6. D.G. Thomas and J.J. Lander, *J. Chem. Phys.* **25**, 1136 (1956).
7. E. Mollwo, *Z. Phys.* **138**, 478 (1954).
8. S.F.J. Cox, E.A. Davis, S.P. Cottrell, P.J.C. King, J.S. Lord, J.M. Gil, H.V. Alberto, R.C. Vilão, J. Piroto Duarte, N. Ayres de Campos, A. Weidinger, R.L. Lichti, and S.J.C. Irvine, *Phys. Rev. Lett.* **86**, 2601 (2001).
9. D.M. Hoffman, A. Hofstaetter, F. Leiter, H. Zhou, F. Henecker, B.K. Meyer, S.B. Orlinskii, J. Schmidt, and P.G. Baranov, *Phys. Rev. Lett.* **88**, 045504 (2002).
10. M.D. McCluskey, S.J. Jokela, K.K. Zhuravlev, P.J. Simpson, and K.G. Lynn, *Appl. Phys. Lett.* **81**, 3807 (2002).
11. S.J. Jokela, M.D. McCluskey, and K.G. Lynn, *Physica B* **340-342**, 221 (2003).
12. N.H. Nickel and K. Fleischer, *Phys. Rev. Lett.* **90**, 197402 (2003).
13. E.V. Lavrov, J. Weber, F. Börrnert, C.G. Van de Walle, R. Helbig, *Phys. Rev. B* **66**, 165205 (2002).
14. Cermet Inc., 1019 Collier Rd., Suite C1, Atlanta, GA 30318.
15. G. Alvin Shi, M. Saboktakin, M. Stavola, and S.J. Pearton (unpublished).

Hydrogen Bonding in ZnO

N. H. Nickel and K. Brendel
Hahn-Meitner-Institut Berlin
Kekuléstr. 5, D-12489 Berlin, Germany.

ABSTRACT

Hydrogen effusion measurements reveal that the total hydrogen concentration in ZnO ranges from 5.2×10^{16} cm^{-3} for single crystal ZnO to 3×10^{21} cm^{-3} for polycrystalline ZnO thin-fims. From the H effusion spectra the hydrogen chemical potential is determined as a function of the H concentration that can be related th the H density-of-states distribution. Single crystal ZnO exhibiting the lowest H concentration reveals six peaks in the H density-of-states located between 0.59 and 1.4 eV below the H transport site. With increasing H concentration the amount of H accommodated with binding energies larger than 1.0 eV increases significantly.

INTRODUCTION

Recently, zinc oxide is attracting a lot of interest because of its electrical and optical properties for a variety of applications ranging from UV light emitting diodes and lasers to piezoelectric devices. However, and major drawback is the fact that the zinc oxide generally exhibits n-type conductivity. Over the last decade it has been established that presence of hydrogen in a semiconductor influences its electrical properties. In most semiconductors H acts as a compensating center neutralizing intentionally incorporated dopants. The behavior of hydrogen is different in zinc oxide. Results obtained from first-principles density-functional calculations suggest that hydrogen acts as a shallow donor independent of the position of the Fermi energy. Therefore, it might be the source of the traditionally observed n-type conductivity in ZnO [1]. This theoretical prediction was confirmed by muon spin rotation measurements [2] and nuclear double-resonance spectroscopy [3]. Hall-effect measurements performed on state-of-the-art single-crystal the zinc oxide revealed to the presence of two donor state [4], one of which was identified as the hydrogen show donor that occurs at a concentration of about 6×10^{16} cm^{3} [3]. In fact, hydrogen effusion measurements performed on undoped state-of-the-art ZnO single crystals revealed a total hydrogen concentration of about 5.2×10^{16} cm^{-3} [5]. This value is an excellent agreement with the observed concentration of hydrogen shallow donors.

In this paper we present hydrogen effusion data obtained from state-of-the-art single crystal ZnO and sputtered zinc oxide thin films. By analyzing the effusion data of the hydrogen density-of-states distribution (H DOS) can be derived. Single-crystal the zinc oxide exhibits six peaks in the H density of states distribution. The most prominent peak is located at about 0.86 eV be low that H transport states. The total hydrogen concentration in the sputter deposited zinc oxide films varies between 10^{20} and 3×10^{21} cm^{-3}. Moreover, the majority of H atoms are accommodated with binding energy exceeding 1.0 eV.

EXPERIMENTAL DETAILS

The single crystal zinc oxide samples used in this study were c-axis oriented crystals with an "O" face and a thickness of 0.5 mm. The ZnO thin-films were doped with 2 % Al. These specimens were grown on quartz substrates by dc magnetron sputtering at a substrate temperature of 150 °C, a power density of 22.6 W/cm^2, and a deposition rate of 2.9 nm/s. The thickness of the samples varied from the 0.11 to 3.83 eV. The ZnO thin-films are composed of columnar grains extending from the substrate to the surface with an average grain diameter of approximately 0.16 µm. The length of the grains is limited only by the film thickness.

To obtain information on the hydrogen concentration and bonding, the samples were annealed in ultra-high vacuum up to 950 °C with a constant heating rate of 20 K/min while the flux of molecular hydrogen was measured with the quadrupole mass spectrometer. Prior to each measurement background spectra were taken and subsequently subtracted from the sample data. To obtain absolute values for the molecular hydrogen flux the data were calibrated with the known neon flux through a capillary.

RESULTS AND DISCUSSION

In Fig. 1 the molecular hydrogen flux of the zinc oxide samples is shown as a function of the temperature T. the lowest hydrogen flux is observed in a state-of-the-art single crystal ZnO sample with the maximum of 1.25×10^{12} cm^{-2}s^{-1} at $T = 473$ °C. The polycrystalline ZnO samples

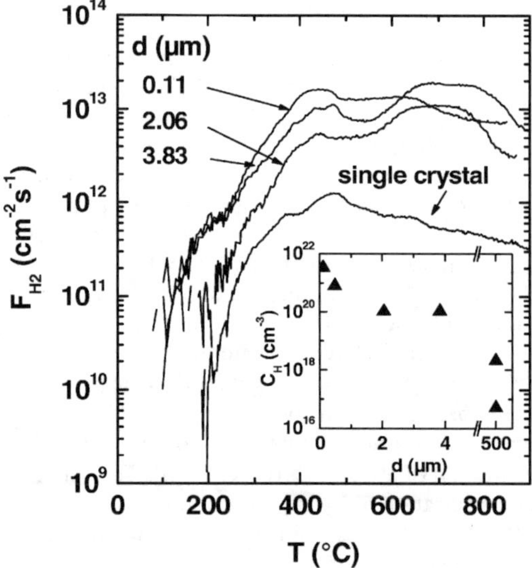

Figure 1. Molecular hydrogen flux of single-crystal ZnO and sputtered deposited ZnO:Al thin films as a function of temperature. The inset shows the total atomic H concentration versus sample thickness. Details are described in the text.

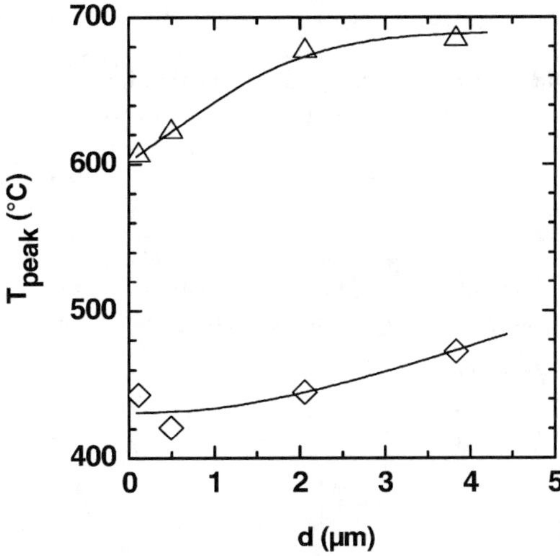

Figure 2. temperature at which the molecular hydrogen flux exhibits a maximum, T_{peak}, as a function of sample thickness.

exhibit a hydrogen flux that exceeds the H_2 flux of the single crystal in the entire temperature range by more than one order of magnitude. In addition, the polycrystalline ZnO samples exhibit a significant hydrogen flux at temperatures below 200 °C. This indicates that the total hydrogen concentration is considerably larger in sputtered zinc oxide samples than in single crystals. The total hydrogen concentration is obtained by integrating the effusion spectra, taking into account the heating rate and the sample thickness. That total hydrogen concentrations of the various samples are plotted as a function of the layer thickness in the inset of Fig. 1. This plot suggests that the H content is distributed inhomogeneously with thickness. This, however, is not the case. The ZnO thin films were deposited consecutively starting with the thinnest sample without breaking the vacuum. Since the largest H concentration of $C_H \approx 3.3 \times 10^{21}$ cm^{-3} is observed in the sample that was deposited first (d = 0.11 µm) and the H concentration decreases in each of the following specimens it is very likely that residual water vapor in the deposition chamber is the source of H. On electron impact H_2O dissociates into H and OH [6,7].

The hydrogen effusion spectra are influenced by the thickness of the ZnO samples. Above 400 °C the effusion spectra reveal two maxima one at low temperatures between 400 and 500 °C and a high temperature maximum between 600 and 700 °C. In Fig. 2 the temperature at which the molecular hydrogen flux exhibits a maximum, T_{peak}, is plotted as a function of the sample thickness. With increasing sample thickness T_{peak} shifts to larger values. For the high and low temperature maxima a shift of 90 °C and 50 °C is observed, respectively, as the sample thickness increases to 3.83 µm. This behavior indicates that the H effusion process in ZnO is governed by the diffusion of monatomic H from the bulk to the surface and the surface defect density does not affect the effusion spectra [8]. At the sample surface the H atoms recombine to H_2, which then is detected by the mass spectrometer. This is also corroborated by the observation that in single crystal ZnO HD molecules are detected at a concentration of approximately 2×10^{16} cm^{-3}, while a signal due to D_2 does not arise [5].

From the hydrogen effusion data the hydrogen density of states distribution can be derived. The total H concentration is related to the density-of-states distribution by the equation

$$C_H = \int_{-\infty}^{\infty} N_H(E) f(T,E,\mu_H)\, dE \qquad (1)$$

where $N_H(E)$ is the density-of-states distribution, $f(T,E,\mu_H)$ is the occupation function, T is the temperature, E is the H binding energy, and μ_H is the hydrogen chemical potential. Partial differentiation yields the H DOS

$$N_H(\mu_H) = \frac{\partial C_H(\mu_H)}{\partial \mu_H} \quad , \qquad (2)$$

The position of the H chemical potential is estimated from the H_2 flux

$$F = F_0 \exp\left(-\frac{E^*-\mu_H}{kT}\right) \quad , \qquad (3)$$

where E^* represents the energy of the H transport sites and k is the Boltzmann constant. The prefactor is given by [9] $d\times F_0 \approx 2a\nu N_{surf}$, where a is the mean free path, ν is an attempt frequency, and N_{surf} is the concentration of H surface states. With reasonable values of $a \approx 3\times10^{-8}$ cm, $\nu \approx 10^{13}$ Hz, and $N_{surf} \approx 10^{15}$ cm^{-3}, a value of $d\times F_0 \approx 6\times10^{20}$ s^{-1}cm^{-1} is obtained. This prefactor represents an average value. However, since F_0 is in the argument of logarithm when computing the H chemical potential even an order of magnitude error in F_0 results in an error of only 0.1 eV in the absolute energy scale.

In Fig. 3 the hydrogen density-of-states distribution for state-of-the-art ZnO single crystals and sputtered ZnO thin films are shown. The lowest H density-of-states distribution is obtained for single crystal ZnO. This is to be expected, since the ZnO single crystal contains the lowest hydrogen concentration (see inset in Fig. 1). Deconvolution of N_H shows that the spectrum is composed of six peaks ranging from 0.59 to 1.4 eV. The most intense peak is located at 0.86 eV. For the sputtered ZnO specimens N_H increases to values ranging from the low 10^{20} cm^{-3} [Fig. 3 (b)] to the mid 10^{21} cm^{-3} [Fig. 3 (d)]. A very important result is shown in Fig 3(b). Both samples contain a H concentration of $\approx 10^{20}$ cm^{-3}. The analysis of the effusion spectra reveals that the H density-of-states is quite similar in both samples independent of thickness. This result clearly demonstrates that the analysis described above is reproducible and reliable.

In single crystal ZnO about 45% of the total hydrogen concentration is accommodated with binding energies less than 1.0 eV. An increase of the H concentration results in an increase of N_H in the energy range 1.0 eV $< E$-$\mu_H < 1.8$ eV for all investigated specimens. On the other hand, the sputtered ZnO samples reveal that only about 25 % of all H atoms form complexes with energies less than 1.0 eV. Most of the hydrogen (≈ 75 %) is accommodated in complexes with binding energies larger than 1.0 eV.

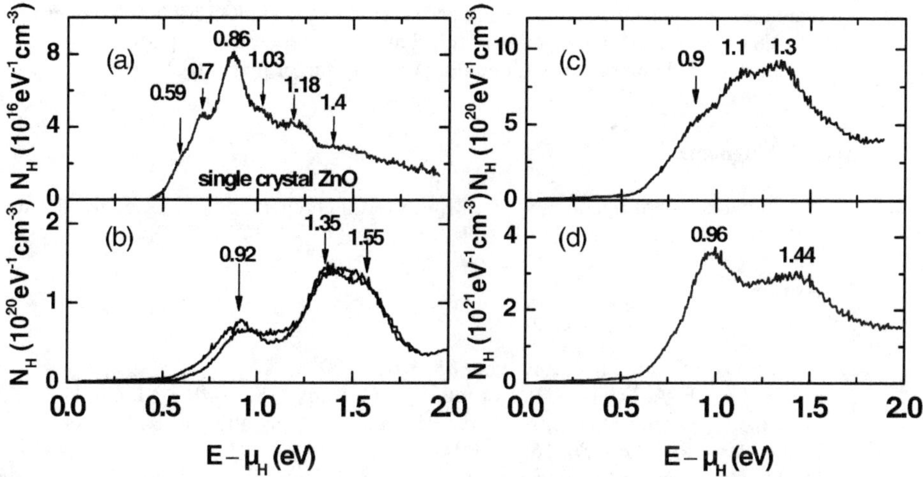

Figure 3. hydrogen density-of-states distribution (N_H vs E-μ_H) in single-crystal zinc oxide (a) and in sputtered ZnO:Al films (b) – (d). The hydrogen density of states is lowest in single-crystal losing oxide. The data were derived from H effusion spectra applying the formalism described in the text.

The most prominent feature in the density-of-states distribution in single crystal ZnO is the peak centered at E-μ_H = 0.86 eV. A possible candidate is a complex consisting of a H atom and an oxygen vacancy. This complex acts as a shallow donor. According to first principles calculations this complex has a binding energy of 0.8 eV [1]. On the other hand, the sputtered ZnO samples reveal a peak at E-μ_H ≈ 0.9 eV. It is very likely that this peak corresponds to the same hydrogen complex as in single crystal ZnO. The somewhat larger binding energy in the sputtered specimens might be due to the presence of strain that is induced by disorder. Tensile strain at the grain boundaries introduces a bond length distortion that can easily account for an increase of the binding energy of the H complex. The required bond length distortion would be rather small to account for the observed increase in binding energy of ≈ 0.05 eV.

In summary, we performed hydrogen effusion measurements on state-of-the-art ZnO single crystals and sputtered Al doped ZnO thin films. Significant H out-diffusion is observed for temperatures above 200 °C. The H effusion mechanism is diffusion limited. The lowest H concentration of is observed in state-of-the-art ZnO single crystals. In polycrystalline ZnO samples the H concentration is up to 5 orders of magnitude higher. The H content depends on the precondition of the preparation chamber; e.g.: the amount of residual water vapor. From the H effusion spectra the hydrogen density-of-states distribution was derived. Single crystal ZnO reveals a H density-of-states distribution that deconvolutes into six peaks. The most prominent peak is located at 0.86 eV. This H binding energy is quite similar to the calculated binding energy for a complex consisting of a H atom and an oxygen vacancy [1]. In sputtered ZnO samples N_H shows a peak approximately at 0.9 eV that might correspond to the same H complex. It is likely that the somewhat larger binding energy is due to bond distortions at grain boundaries.

An increase in H concentration results in an increase of the H DOS for binding energies between 1.0 and 1.8 eV. Approximately 75 % of the H atoms incorporated in sputtered ZnO are accommodated in H complexes with binding energies larger than 1.0 eV.

Acknowledgment

The authors would like to thank C. Klimm for the preparation of the sputtered ZnO samples.

References

1. C. G. Van de Walle, Phys. Rev. Lett. **85**, 1012 (2000).
2. S. F. J. Cox, E. A. Davis, S. P. Cottrell, P. J. C. King, J. S. Lord, J. M. Gil, H. V. Alberto, R. C. Vilao, J. Pironto Duarte, N. Ayres de Campos, A. Weidinger, R. L. Lichti, and S. J. C. Irvine, Phys. Rev. Lett. **86**, 2601 (2001).
3. D. M. Hofmann, A. Hofstaetter, F. Leiter, H. Zhou, F. Henecker, B. K. Meyer, S. B. Orlinskii, J. Schmidt, and P. G. Baranov, Phys. Rev. Lett. **88**, 045504 (2002).
4. D. C. Look, D. C. Reynolds, J. R. Sizelove, R. L. Jones, C. W. Litton, G. Cantwell, and W. C. Harsch, Solid State Commun. **105**, 399 (1998).
5. N. H. Nickel and K. Fleischer, Phys. Rev. Lett. 90, 197402 (2003).
6. J. W. Bozzelli and R. B. Barat, Plasma Chem. Plasma Process. 8, 293 (1983).
7. S. Roychowdhury, U. K. Roychowdhury, and M. Venugopalan, Plasma Chem. Plasma Process. 2, (1982).
8. W. Beyer, in Hydrogen in Semiconductors II, edited by N. H. Nickel (Academic Press, San Diego, 1999), Vol. 61, p. 165.
9. W. B. Jackson, A. J. Franz, H.-C. Jin, J. R. Abelson, and J. L. Gland, J. Non-Cryst. Sol. 227-230, 143 (1998).

Characteristic of hydrogenated Ga-doped ZnO films grown by DC magnetron sputtering using H_2/Ar gas

Satoshi Takeda and Makoto Fukawa
Research Center, Asahi Glass Co., Ltd.
1150 Hazawa-cho, Kanagawa-ku, Yokohama 221-8755, Japan

ABSTRACT

We report characteristics of sputtered Ga-doped ZnO films grown in pure Ar (Ar-films) and H_2/Ar gas mixture (H_2/Ar-films). With increasing water partial pressure (P_{H2O}), the resistivity of Ar-films significantly increased due to the decrease in both free carrier density and Hall mobility. The transmittance in the wavelength region of 300-400 nm for the films also increased with increasing P_{H2O}. However, no significant P_{H2O} dependence of the electrical and optical properties was observed for H_2/Ar-films. Secondary ion mass spectrometry analysis revealed that hydrogen concentration in the Ar-films increased with increasing P_{H2O}, indicating that the origin of the incorporated hydrogen is attributed to the residual water vapor in the coating chamber. On the contrary, the hydrogen concentration in H_2/Ar-films was almost constant irrespective of P_{H2O} and the concentration is obviously higher than that of Ar-films, indicating that the hydrogen primarily comes from H_2 gas and the adsorption species due to H_2 gas preferentially adsorb to the growing film surface over residual water vapor. Consequently, the effect of P_{H2O} on the crystal growth was significantly reduced by H_2 gas introduction, resulting in stability improvement.

INTRODUCTION

Hydrogen is the smallest and lightest element, so that it easily diffuses into materials. The diffused hydrogen can have effects on the material properties [1,2]. Therefore, it is important to clarify the role of hydrogen to control the material properties. Secondary ion mass spectrometry (SIMS) is widely used to detect hydrogen because of its excellent sensitivity and high depth resolution. Also, an isotope analysis is possible in this method, so that SIMS analysis using deuterium (D_2) is considered to be useful to elucidate the origin of hydrogen.

In the present study, we applied the SIMS analysis to investigate the hydrogen in Ga-doped ZnO (GZO) films grown by DC magnetron sputtering. GZO is a typical transparent and conductive oxide, which is degenerate wide band-gap semiconductors with low resistance and high transparency in the visible wavelength range, and is one of the promising materials in optoelectronic devices such as transparent electrodes in flat panel displays and solar cells [3,4]. Here, GZO films were deposited under pure Ar and H_2/Ar mixture gas. The effects of H_2 gas introduction on the stability of both electrical and optical properties of the films versus residual water pressures (P_{H2O}) were investigated in detail.

EXPERIMENTAL DETAILS

The GZO films were deposited onto glass (Corning 7059) with a thickness of 150-180nm by DC magnetron sputtering at room temperature under various residual water pressures (P_{H2O}). Quadrupole mass spectrometry analysis revealed that base pressure was primarily due to P_{H2O}. Thus, P_{H2O} were controlled by the base pressure in this study. The sputtering was carried out under a total gas pressure of 1.3 Pa of Ar 100 %, $H_2/Ar=1/5$ and deuterium (D_2)/Ar=1/5.

Observation of surface morphology of the films was performed using a SEM. Crystalline phases of the films were identified by glancing-angle XRD using Cu-Kα radiation operated at 50 kV–200 mA with the incident angle of 0.5°. Hydrogen and deuterium in the films were measured by SIMS. The secondary ions of $^1H^-$ and $^2D^-$ were detected using a 5 keV Cs^+ primary ion beam with a beam current of 200 nA. The angle of incidence was 60° to the normal of the sample surface. The optical transmission and reflection spectra of the films were measured at room temperature in air using a dual beam spectrometer. The resistivity (ρ), Hall mobility (μ) and free carrier density (n) were estimated by the four-point probe method and Hall-effect measurement in the van der Pauw method.

RESULTS AND DISCUSSION

Electrical and optical properties

Figure 1 shows the resistivity (ρ), carrier density (n) and Hall mobility (μ) of GZO films deposited in Ar 100% (Ar-films) or H_2/Ar gas mixture (H_2/Ar-films) as a function of P_{H2O}. With increasing P_{H2O}, the ρ of Ar-films increases due to the decrease in both n and μ. On the other hand, no significant change in ρ is observed for H_2/Ar-films. These results indicate that the introduction of H_2 gas significantly improves stability of electrical properties of the films against trace water vapor in the coating chamber.

Figure 2 shows the transmission spectra of Ar and H_2/Ar-films deposited under various P_{H2O}. The transmittance in the wavelength region of 300-400 nm increases with increasing P_{H2O} for the Ar-films. This phenomenon is not observed for the H_2/Ar-films. These results indicate that the introduction of H_2 gas is an effective way to improve the stability of both electrical and optical properties of the films versus P_{H2O}. Similar tendency was observed by the introduction of D_2 gas.

Origin of hydrogen

Figure 3 shows the secondary ion intensity ratio ($^1H^-/^{80}ZnO^-$) which represents the hydrogen concentration in the films as a function of P_{H2O}. Here, we focus on the origin of hydrogen because information about chemical states of hydrogen cannot be obtained in SIMS analysis. With increasing P_{H2O}, the hydrogen concentration in Ar-films increases, indicating that the origin of the hydrogen is due to residual water in the coating chamber. On the other hand, the hydrogen concentration in H_2/Ar-films is almost constant irrespective of P_{H2O} and the concentration is obviously higher than that of Ar-films even when P_{H2O} is increased up to 4.0×10^{-3} Pa. It was also revealed that $^2D^-$ was clearly incorporated into D_2/Ar-films. These results indicate that the hydrogen in the H_2/Ar-films

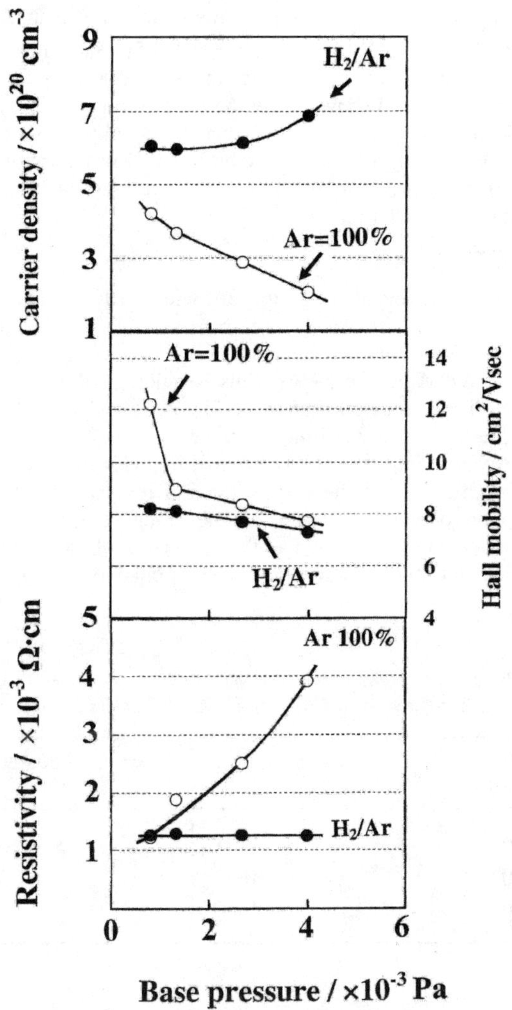

Figure 1. The resistivity (ρ), carrier density (n) and Hall mobility (μ) for GZO films deposited in Ar 100% or H_2/Ar gas mixture as a function of P_{H2O}.

primarily comes from H_2 gas. It was confirmed that the hydrogen was uniformly distributed into both Ar- and H_2/Ar-films from $^1H^-$ depth profiles. This fact indicates that some species including hydrogen is continuously supplied to the growing film surface during the deposition, that is, the film surface is covered by the adsorption species due to H_2O or H_2. Taking these into consideration, it is considered that the adsorption species from H_2 gas preferentially adsorb to the growing film surface over residual water vapor in H_2/Ar plasma process. Consequently, the stability of electrical and optical properties of H_2/Ar-films versus P_{H2O} is significantly improved.

Film structure

X-ray diffraction analysis revealed that the full width at half maximum (FWHM) of ZnO (002) peak for the Ar-films increased with increasing P_{H2O}, as shown in Table I. This result indicates that grain size of the films decreases with increasing P_{H2O}. On the contrary, the degree of change in FWHM for the H_2/Ar-films is much smaller than that of Ar-films although the values of FWHM are larger than those of Ar-films. These observations clearly indicate that the grain size of the H_2/Ar-films is smaller than that of Ar-films, but that the degree of change in the grain size of the H_2/Ar-films along with P_{H2O} is much smaller than that of the Ar-films. On the basis of these experimental results, it is concluded that the variation of electrical and optical properties of the films results from the change in microstructure of the films along with P_{H2O}, and that the origin of the stability improvement by H_2 gas introduction is due to the reduction of degree of change in microstructure.

Table I The full width at half maximum (FWHM) of ZnO (002) peak obtained by XRD measurements for GZO films deposited in Ar 100 % or H_2/Ar=1/5 gas under various P_{H2O}.

P_{H2O} / Pa	FWHM / degree	
	Ar-films	H_2/Ar-film
4.0×10^{-3}	0.71	0.99
1.3×10^{-3}	0.57	0.93
8.0×10^{-4}	0.50	0.91

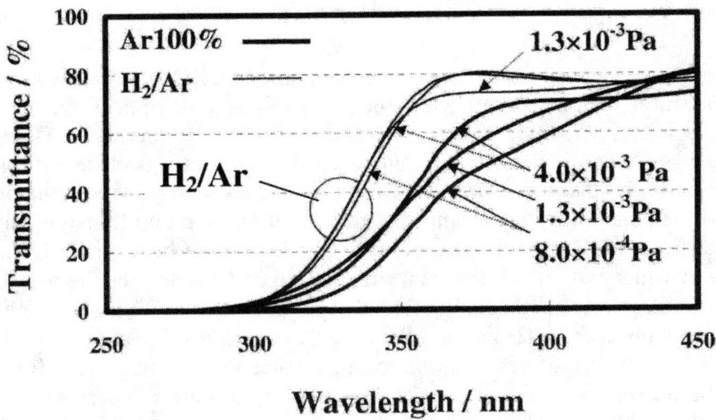

Figure 2. The transmission spectra of GZO films deposited in Ar 100 % or H_2/Ar gas mixture under various P_{H2O}.

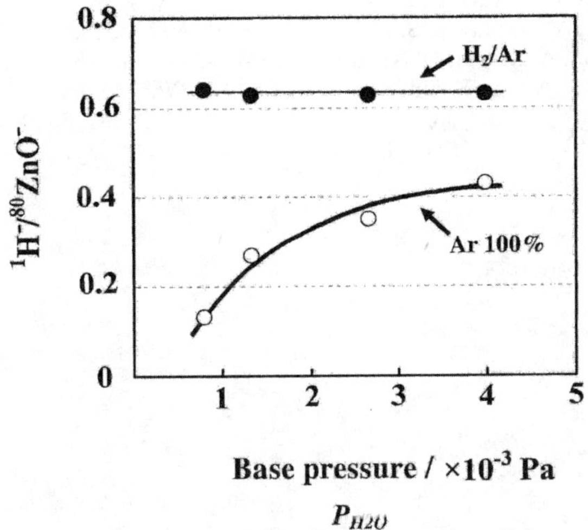

Figure 3. The secondary ion intensity ratio ($^1H^-/^{80}ZnO^-$) for GZO films deposited in Ar 100% or H_2/Ar gas mixture as a function of P_{H2O}.

CONCLUSIONS

In this paper, we have reported characteristic of sputtered GZO films grown in pure Ar and H_2/Ar gas mixture. With increasing P_{H2O}, the resistivity and the optical transmittance in the wavelength region of 300-400 nm of the Ar-films significantly increased. However, no significant P_{H2O} dependence of the electrical and optical properties was observed for the H_2/Ar-films. It was found that grain size of Ar-films decreases with increasing incorporated hydrogen due to residual water vapor, and that the variation of ρ and transmittance along with P_{H2O} of the films resulted from change in the grain size. However, the incorporated hydrogen of H_2/Ar-films was almost constant irrespective of P_{H2O} and the degree of change in grain size of the films was much smaller than that of the Ar-films. These indicate that the hydrogen primarily comes from H_2 gas and the adsorption species due to H_2 gas adsorbs to the growing film surface in preference to the residual water vapor. That is, the film surface is covered by the adsorption species from H_2 gas during the deposition in H_2/Ar plasma process, so that the stability of electrical and optical properties versus P_{H2O} is significantly improved. These results may serve as clues on how to modify film properties using hydrogen gas.

REFERENCES

1. J. I. Pankove, *Phy. Rev. Lett.* **51**, 2224 (1983).
2. Y. Yasuda, Y. Koide, A. Furukawa, S. Zaima, *J. Appl. Phys.* **73**, 2288 (1993).
3. M. Miyazaki, S. Sato, A. Mitsui, H. Nichimura, *J. Non-Cryst. Solids* **218**, 323 (1997).
4. M. Kon, P.K. Song, A. Mitsui, Y. Shigesato, *Jpn. J. Appl. Phys.* **41**, 6174 (2002).

Hydrogen Intercalation Sites in Rutile Predicted from Ab Initio Calculations.

Marina V. Koudriachova, Nicholas M. Harrison and Simon W. de Leeuw

Physical Chemistry and Molecular Thermodynamics, DelftChemTech, Delft University of Technology, Julianalaan 136, 2628 BL Delft, the Netherlands
Department of Chemistry, Imperial College of Science and Technology, London, SW7 2AZ, UK
and CCLRC, Daresbury Laboratory, Daresbury, Warrington WA4 4AD, UK

Abstract

The geometry of hydrogen intercalation sites in rutile TiO_2 has been examined with first principles calculations. The calculations predict a new intercalation site, which is consistent with current experimental data. The apparent contradictions between spectroscopic, diffraction and magnetic resonance data are resolved by taking into account the strong local distortions of the structure induced by localization of charge donated by hydrogen. In particular the predicted OH vibrational frequency is in excellent agreement with the experimentally measured value. A similar local geometry is predicted in the presence of trivalent counter ions but with a preferential site occupancy consistent with the local symmetry implied by spectroscopic measurements of aluminium substituted titania.

Introduction

The large and growing use of nanostructured titania in solid state devices such as solar cells, battery electrodes has lead to a great deal of interest in the influence of hydrogen impurities on their properties. The presence of hydrogen ions strongly affects electronic, optical and structural properties of the host material. The strong charge transfer from hydrogen modifies its electronic structure. At elevated temperatures, proton transport makes a significant contribution to the ionic conductivity.

From a technological point f view rutile structured TiO_2 is the most important polymorph of titiania. Natural samples contain small amounts of hydrogen which is thought play a role in the charge compensation of trivalent cation impurities. Hydrogen may play a similar role in commercial rutile pigments, which are intentionally doped with Al^{3+} cations to suppress undesirable photoactivity. Hydrogen can be introduced reversibly into rutile by annealing in an atmosphere containing H_2O+O_2 or H_2 up to concentrations[1] 10^{19} /cm^3 . Intercalation is thought to proceed through dissociative adsorption at surface defects[2,3] although there is some evidence for molecular intercalation in the near surface region[4].

Rutile TiO_2 has a tetragonal structure, consisting of TiO_6 octahedra which share edges along the c-direction and corners in the ab-planes, so that the pillars of edge sharing octahedra along the c-direction are separated in the ab-planes by channels of empty octahedral, as shown in figure 1.

Information about the sites adopted by hydrogen impurities can be obtained from the clearly differentiated O-H vibrational mode around 3277 cm^{-1} through analysis of infra-red (IR) adsorption and Raman data. The large dichroism observed in the IR adsorption and symmetry analysis of the polarised Raman spectra indicate that the OH-bonds are oriented perpendicular to the tetragonal *c*-axis of rutile[2,5,6]. The two types of intercalation sites are consistent with this orientation. In the channel centre (CC) model[2] a proton resides in the channel on a site equidistant between two O^{2-}-ions separated by 3.33 Å (Figure 1). In the modified CC-model (MCC)) the proton is shifted towards one of the two O^{2-}-ions forming an OH bond with a bondlength of 0.98 Å[2]. In the BOE model[7] the proton resides on a site equidistant from the two O^{2-}-ions which make up the shared (basal) octahedron edge along either (110) or the (1$\overline{1}$0) directions. These oxide-ions are separated by 2.85 Å (Figure 1).

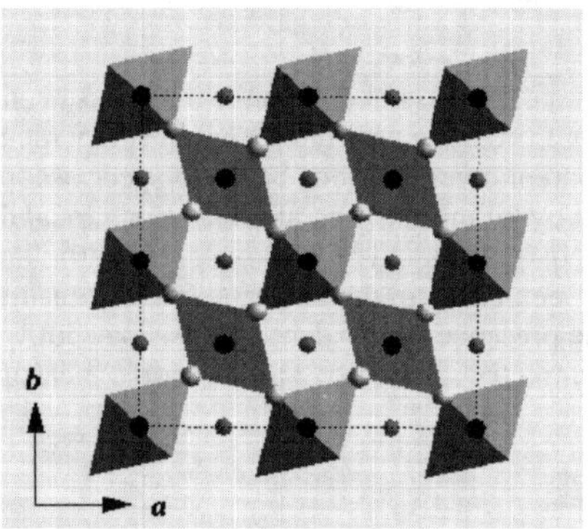

Figure1. Eight unit cells of rutile TiO_2 viewed along the crystallographic *c*-direction. Two symmetry distinct H adsorption sites are indicated; the CC site in grey, the BOE-site in black.

In pure rutile symmetry analysis of the Raman data excludes the BOE mode and is consistent with the CC model[5,6] - the diagonal σ(110) mirror plane implied by the BOE model is not consistent with the Raman data. In rutile doped with Al^{3+}-ions this symmetry element is present so that BOE model may pertain in this case[6]. At the same time EPR measurements of rutile doped with another trivalent cation, Fe^{3+}, support the CC model[8]. The geometry of neither CC (MCC) nor BOE models is consistent with the observed vibrational frequency of the OH group. There is an empirical relation between the O-H stretch frequency and the O-O distance of an O-H-O complex[9]. The O-O distance in the MCC model is 3.33 Å, for which a frequency of 3600 cm^{-1} is predicted. This frequency is almost that of a free O-H group. In the BOE model the O-O is 2.52 Å and the vibrational frequency should be shifted downwards to 2300 cm^{-1}. The observed OH frequency of 3279 cm^{-1} correlates with an O-O distance of 2.75-2.85 Å instead.

This paper discusses the results of first principles simulations based on density functional theory applied to resolve this anomaly. The calculations predict the stable hydrogen sites upon intercalations in pure rutile and that doped with trivalent cations. It will be shown that the apparent contradiction between the spectroscopic data and local geometry is resolved once the local deformations of the lattice are included. These deformations are caused by accommodation of the electron density donated by hydrogen upon intercalation. Currently available experimental data are discussed in the light of these results.

Details of calculations

All calculations were performed with the CASTEP software within the pseudopotential plane-wave formalism[10,11] and density functional theory. Electron exchange and correlation effects were treated within the spin-polarized generalized gradient approximation[12] with ultrasoft pseudopotentials[13] used to replace the Ti(1s,2s,2p), O(1s) and H(1s) orbitals. The sampling in reciprocal space was performed on a regular grid with a spacing in the of 0.1 Å^{-1}. A plane wave cutoff energy of 380 eV was found to converge the total energy to 0.01 eV per formula unit. The calculations were performed in a supercell containing 16 formula units of TiO_2 and one hydrogen atom. The size and shape of the cell and all internal degrees of freedom were fully relaxed with respect to the total energy. The distribution of electron density was monitored using the Mulliken population analysis.

Results and discussion

The structural parameters of rutile are predicted in excellent agreement with experimental values: a=b=4.59 Å and c=2.95 Å[14], calculated values a=b=4.63 Å and c=2.96 Å. A hydrogen atom was inserted into the BOE and MCC sites respectively and the structures were fully relaxed. The BOE structures were found to be unstable with respect to the CC model. Upon relaxation H^+-ions leave the edge of the octahedron and move into the c-channel, as shown in Figure 2. H^+-ions occupy sites which are symmetrically equivalent to (0.54a, 0.09b, 0c) of the primitive cell of rutile. These are similar to the sites deduced from EPR measurements in Fe-doped rutile ((0.56a, 0.11b, 0c) with Fe-ions at (0.5a, 0.5b, 0.5c))[8]. In these sites the hydroxyl group has a bond length of 1.00 Å and is oriented close to either the (110) or ($1\bar{1}0$) directions. There is considerable local distortion of the lattice which results in the hydrogen bond distance across the c-channel being 1.86 Å rather than 2.35 Å of the CC model. This distortion immediately removes the main difficulty of the CC model discussed above. The frequency obtained by fitting the computed energy surface for the O-H bond is 3280 cm^{-1}, in excellent agreement with the experimentally observed value of 3288 cm^{-1} at 10 K [2].

The size and the shape of the c-channels are modified significantly by the presence of H^+-ions. Recently it has been shown that the structural expansion of rutile upon insertion of small ions, such as H^+ and Li^+, can be fully rationalized in terms of accommodation of charge donated to the host lattice[15]. In case of hydrogen insertion each hydrogen atom donates $0.32e$ to the host lattice. This charge is strongly localized on the oxide-ions forming the hydroxyl group (about 28 %) and on oxide-ions across the c-channel forming hydrogen bonds (about 12 %). The remaining charge is accommodated by neighbouring Ti-ions in the *ab* planes and neighbouring oxide ions in the c-direction. The charge donated to oxide and titanium ions increases their ionic radius by a considerable amount with a concomitant elongation, by 7 %, of the Ti-O bonds involved.

This reshapes the c-channel, as it is shown in Figure 2. The c-channel becomes more tight in the direction of the O-H-O complex (for instance, (110)). At the same time the channel becomes more spacious in a perpendicular direction (($1\bar{1}0$)). The corresponding atomic displacement pattern is reminiscent to that of the low frequency B_{1g} mode[16] at 142 cm^{-1}.

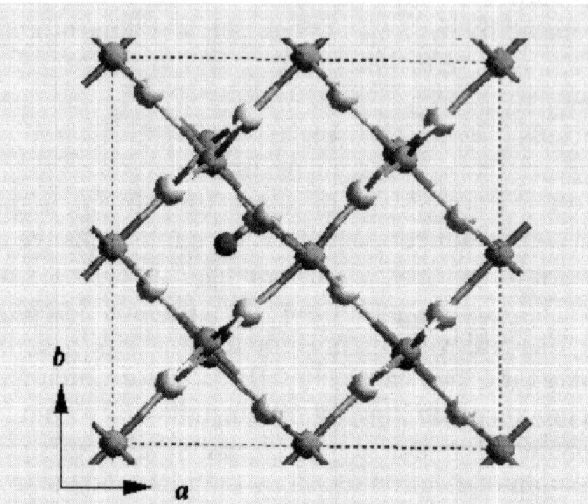

Figure 2. Eight unit cells of rutile containing a H$^+$-ion viewed along the c-direction.. Ti^{4+} ions are shown in light grey, O^{2-} -ions in white, the H$^+$-ion in dark grey.

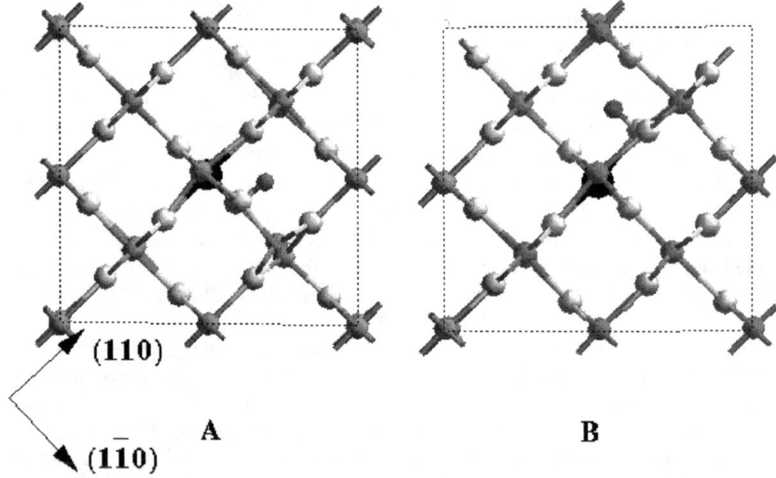

Figure 3. Eight unit cells of rutile doped with Al-ions containing a H$^+$-ion viewed along the c-direction. Ti^{4+} ions are shown in light grey, O^{2-} -ions in white, the Al^{3+}-ion in black and a H$^+$-ion in dark grey. The pictures show H-bonding to O^{2-}-ions at the shared edge of the Ti and Al octahedra (A) and at the shared corner (B).

Upon substitution of a Ti^{4+}-ion by a trivalent ion (Fe, Al) at (0.5a, 0.5b, 0.5c) the stable configurations for intercalated H^+-ions remain locally similar to the distorted CC structure. The computed position of the H^+-ion for the case of Al substitution is (0.54a,0.12b,0c). To our knowledge this has not been determined experimentally. In the Fe case the computed position of (0.54a, 0.10b, 0c) is in excellent agreement with that deduced from EPR measurements (0.56a, 0.11b, 0c[13]).

The substitution and subsequent relaxation creates preferred binding sites for hydrogen: oxide ions at the shared edge of Ti and Al octahedra. In these sites the orientation of the OH-bonds is close to the (110) direction rather than along both the $(1\overline{1}0)$ and (110) directions as in pure rutile. Bonding at the (110) oriented site is preferred over bonding at the $(1\overline{1}0)$ oriented sites: by 0.22 eV in Al and 0.4 eV in Fe-substituted rutile TiO_2. This preferential site occupancy restores a $\sigma(110)$ mirror plane, whose presence was previously interpreted as evidence for the BOE model[11].

The preference of the H^+-ions for bonding at the shared edge of Ti and Al octahedra is largely electrostatic in origin. The Al^{3+}-ion is more electronegative than the Ti^{4+}-ions and thus its nearest neighboring Ti-ions (which are along the c-directions) are partly oxidized having charges of 1.41|e| compared to 1.36-1.37|e| on other Ti-ions. The O^{2-}-ions of the shared edge also lose some of their charge to the Al^{3+} ion (charge -0.74|e| compared to -0.78|e| on the neighboring O^{2-}-ions co-planar with the Al^{3+}-ions in the ab-planes). Thus the intercalated H, as an electron donor, is stabilized at the shared edge as it donates electron density to these charge deficient ions. The computed O-H-O distance is 3.13 Å for Al-doped rutile, significantly longer than that for pure rutile (2.86 Å) and this corresponds to a higher OH frequency as observed in IR and Raman experiments (3331-3325 cm^{-1} on Al-doping[1,5]).

Conclusions

The hydrogen intercalation sites in rutile TiO_2 have been determined using first principles calculations. The predicted site is similar to the channel center model with strong local lattice distortions which locally reshapes the channels. This removes all inconsistencies between this model and existing experimental data. The primary cause of the distortions is found to be the localization of the charge donated by hydrogen upon insertion.

Upon doping with trivalent cations the channel center position is retained but preferential site occupancy leads to a different symmetry for the OH-stretching vibration which was previously interpreted as evidence for the BOE model. Local deformations of the lattice upon substitution are also predicted to result in a frequency shift of the OH stretching vibration which is calculated in excellent agreement with spectroscopic measurements.

References

1. O.W. Johnson, W.D. Ohlsen and P.I. Kingsbury, Phys. Rev. **175**, 1102 (1968)

2. W. Göpel, G. Rocker and R. Feierabend, Phys. Rev **B 28**, 3427 (1983)

3. J.-M. Pan, B.L. Maschhoff, U. Diebold and T.F. Madeym J. Vac. Sci. Technol. **A10**, 2470 (1992)

4. J.V. Cathcart, R.A. Perkins, J.B. Bates and L.C. Manley,   J. Appl. Phys.**50**, 4110 (1979)

5. S. Klauer and M. Wöhlecke, Europhys. Lett. **20**, 439 (1992)

6. S. Klauer and M. Wöhlecke, Phys. Rev. **49**, 158 (1994)

7. A. von Hippel, J. Kalnajas, and W.B. Westphal, J.Phys. Chem.Solids **23**, 779 (1962)

8. P.O. Anderson, E.L. Kollberg and A. Jelenski, Phys. Rev. **B8**, 4956 (1973)

9. G. C. Pimentel and A. L. McClellan, The Hydrogen Bond (Freeman, San Francisco, 1960), Chapter 3.

10. M.C. Payne, M.P. Teter, D.C. Allan, T.A. Arias and J.D. Joannapoulos, Rev. Mod.Phys. **64**, 1045 (1992)

11. CASTEP 3.9 Academic version, licensed under the UKCP-MSI agreement, 1999

12. J.P. Perdew, Phys. Rev. **B34**, 7406(E) (1986)

13. D. Vanderbilt, Phys. Rev. **B41**, 7892 (1990)

14. J.K. Burdet, T. Hughbanks, G.J. Miller, J.W. Richardson and J.V. Smith, J. Am. Chem. Soc.**109**, 3639 (1987)

15.M.V. Koudriachova, N.M. Harrison and S.W. de Leeuw, Phys. Rev. **B65,** 235423 (2002)

16. C. Lee, P. Ghosez and X. Gonze, Phys. Rev. **B50**, 13379 (1994)

Effect of Hydrogen Treatment on Room -Temperature Electric-Field Induced Properties in Narrow-Gap ZnCdHgTe Thin Films

Galina M. Khlyap and Petro G. Sydorchuk
State Pedagogical University, 24 Franko str., Drohobych, 82100, Ukraine
Jacek Polit[1]
[1]Institute of Physics, University of Rzeszow, 35-310 Rzeszow, Rejtana 16A, Poland

ABSTRACT

The effect of hydrogen treatment on room temperature electric properties of narrow-gap semiconductor thin films $Zn_xCd_yHg_{1-z-y}Te$ ($0 < x < 0.50, 0.20 < y < 0.40$) is investigated for the first time. ZnCdHgTe films of $2 - 5$ μm thickness were grown on glass substrates by pulsed laser deposition technique. As-grown films were thermally treated in the flow of molecular H_2 at 200^0C during 24 hours. Comparison between electric characteristics measured before and after hydrogenation showed sufficient changes of the film resistance and appearance of photosensitivity in the visible wavelength range. Study of current-voltage characteristics of the films revealed appearance and significant change of diode-like properties.

INTRODUCTION

The narrow-gap semiconductor solid solution ZnCdHgTe has been investigated for decades as an alternative material to well-known HgCdTe [1 and references therein]. Requirements of the large area thin film-based optoelectronics force one to make efforts for simple and reliable obtaining active elements with in-advance determined properties. Pulsed laser deposition (PLD) technique is seemed to be one of the requested methods.

Recently [1] we have reported electric characteristics of heterostructures based on PLD-grown $Zn_xCd_yHg_{1-x-y}Te$ ($0 < x < 0.5, 0 < y < 0.5$) films. The main problem of practical applicability of as-grown films is how to make them photosensitive. The samples subjected to the study were of p-type conductivity. Annealing in mercury atmosphere inversed the conductivity type, but caused no registerable photosensitivity. So, the aim of the experiments presented in the paper is to show the possibility of inducing photosensitivity by means of heating samples in the flow of molecular hydrogen. In spite of numerous publications considering effect of H_2 plasma on different properties of the semiconductor materials [2 4] there are no references concerning the influence of molecular hydrogen flow on the properties of A^2B^6 compounds (bulk as well as films).

EXPERIMENTAL DETAILS

As it was described previously [1], ZnCdHgTe films were grown by means of PLD using laser parameters listed in the Table 1.

Table 1. Parameters of Nd:YAG laser used for growth of investigated heterostructures.

Laser	Wavelength, μm	Pulse duration, μs	Repetition rate, Hz	Pulse peak power, J
Nd:YAG	1.06	150	10	0.6

X-ray diffraction studies of the as-grown samples (Fig.1) have shown the single crystal structure of the layers. Microstructure analysis revealed the following composition of the films: x = 0.08, y = 0.37 (further the samples will be referred as Set 1) and x = 0.48, y = 0.22 (further referred as Set 2). Room temperature electric measurements and estimations of intrinsic carrier concentration n_i and gap E_g were performed according to expressions (1)-(2) [1] and summarized in the Table 1.

$$n_i(x, y, T) = \left[6.48 - 4.42x - 6.54y + 1.42 \times 10^{-3}(1 + x + y)\right]10^{14} E_g^{3/4} T^{3/2} \exp\left(-\frac{E_g}{2 k_B T}\right) \quad (1)$$

$$E_g(x, y, T) = -0.301 + 1.93x + 2.29 \times 10^{-2} y^{1/2} + 2.731y - 1.62 x^2 + 5.35 \times 10^{-4} T(1 - 2x - 0.35 y^{1/2} - 1.28y) + 3.328 x^3 - 1.248 y^2 + 2.132 y^3 \quad (2)$$

Table 2. Parameters of the as-grown films.

Composition	Gap width E_g, eV	Intrinsic carrier concentration n_i, cm^{-3}	Resistance ($\Omega \cdot$cm)
x = 0.08, y = 0.37	0.86	$1.1 \cdot 10^{11}$	$2.0 \cdot 10^7$
x = 0.48, y = 0.22	1.13	$2.5 \cdot 10^8$	$2.0 \cdot 10^6$

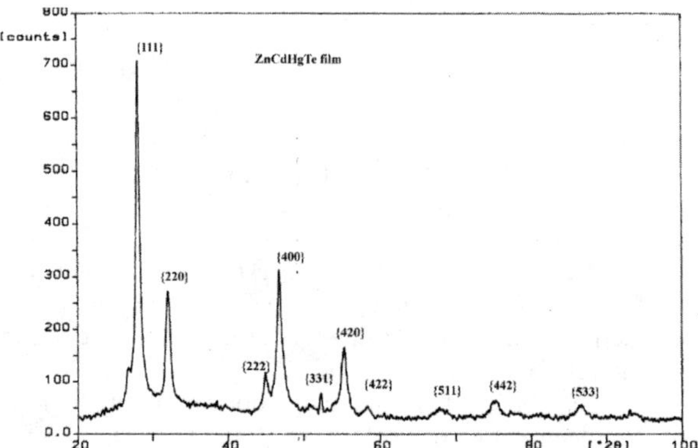

Figure 1. X-ray diffraction spectrum of the as-grown ZnCdHgTe-layer.

For further experiments we selected the best samples with electric area of 1 cm^2. The as-grown layers were treated by means of very simple setup (Fig.2) in the flow of molecular hydrogen under 200^{0}C and pressure P = 10^{-4} Torr during 24 hours. Vacuum valves V1 – V3 have supplied a steady vacuum level monitored by the vacuum controller 2. The ampoule 5 contained the sample was twice evacuated in order to maximally avoid atmospheric oxygen impurity. After finishing the evacuation process and monitoring the concentration of atmospheric oxygen by means of the standard gas controlling equipment (the description is not subjected to the paper) dry hydrogen filled the ampoule. Heat treatment of the evacuated volume was provided by specially constructed electric heater (Fig.2, down panel) supplied the corresponding temperature gradient. In 24 hours, heating was finished and after cooling the ampoule the sample was prepared for electrical measurements.

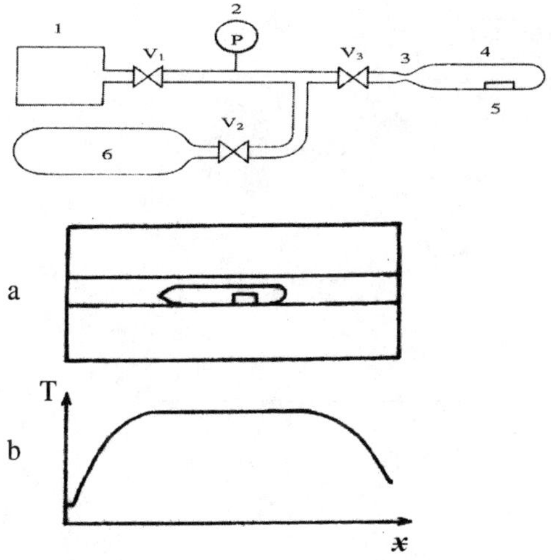

Figure 2. Experimental setup for hydrogenation of ZnCdHgTe films (upper panel): V1, V2 and V3 are vacuum valves, 1 is a vacuum pump, 2 is the vacuum level controller, 3, 4 stand for the evacuated ampoule, 5 takes care of the sample, 6 presents dry hydrogen reserve tube. The lower panel plots the electric furnace (a) and temperature distribution provided by the device (b).

RESULTS AND DISCUSSION

As it was mentioned above, two sets of the samples presented films of different gap were chosen (see Table 2). Electric measurements were carried out under applied electric fields not exceeded 400 V/m (corresponding to the applied voltage up to 3 V). Data of room temperature current-voltage characteristics (IVC) are plotted in the Fig. 3.

As it is shown, the sample Set No 2 (Fig. 3a) demonstrates oscillation-like IVC with no special features. On the contrary, the IVC of the sample Set No 2 (Fig. 3b) under small bias (up to 1 V, the range of particular interest for device applications) can be described by the Fowler – Nordheim function $j/V_a^2 = f(V_a^{-1})$ [5], where V_a is an applied voltage:

$$j \sim V_a^2 \exp(-const\sqrt{\phi_b}), \qquad (1)$$

where ϕ_b takes care of the barrier height, *const* is a function strongly depending on the parameters of the material.. Numerical estimation gives $\phi_b \sim 1.43$ eV. As the applied voltage increases, the rate of current increase becomes smaller and in the range of bias from 1.2 to 3.0 V the current plateau is observed.

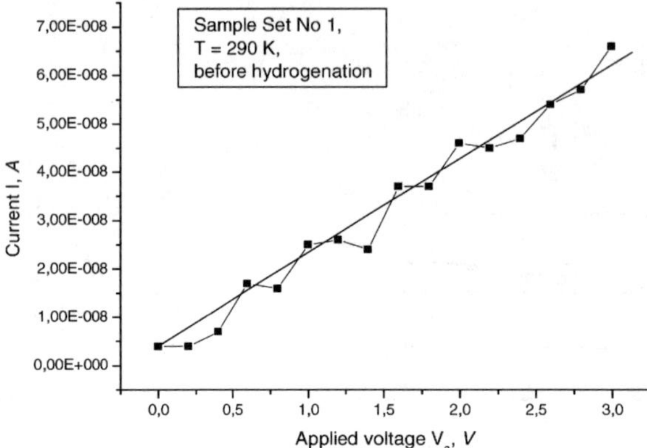

Figure 3, a.

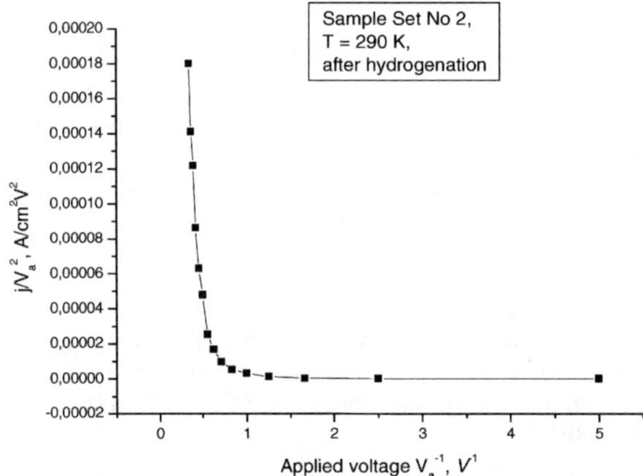

Figure 3, b.

Figure 3. IVC's of the investigated samples before (Fig.3a, solid line is a guide for the eye) and after (Fig.3b) hydrogenation.

Results obtained after hydrogenation of the samples are shown in Fig. 4. It is obvious that the dependencies are sufficiently changed.

The sample Set No 1 (Fig. 4a) demonstrates two linear sections: the diffusion current under the small applied voltage (up to 800 meV) is replaced by the space charge limited (SCL) current described as follows [6]:

$$I = T_{tun} \frac{9\varepsilon\,\varepsilon_0 \mu A}{8L^3} V_a^2, \quad (2)$$

where T_{tun} is the tunneling transparency coefficient (calculated according to [5] and not subjected to the paper), $\varepsilon = 15$ is the dielectric constant of ZnCdHgTe, μ is the carriers mobility estimated to be about 1000 cm^2/Vs. This result indicates a considerable improvement of the film quality.

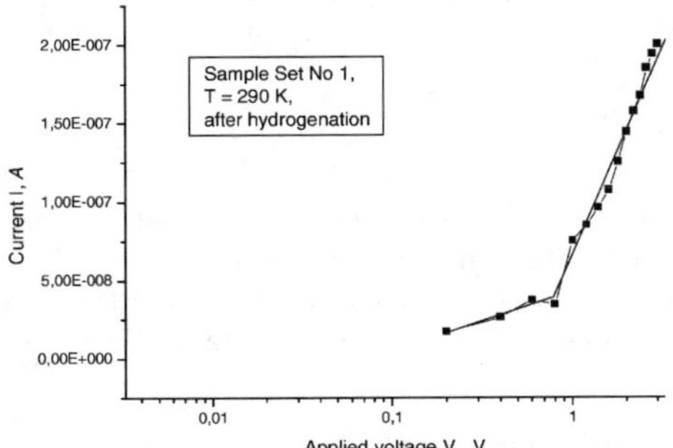

Figure 4a.

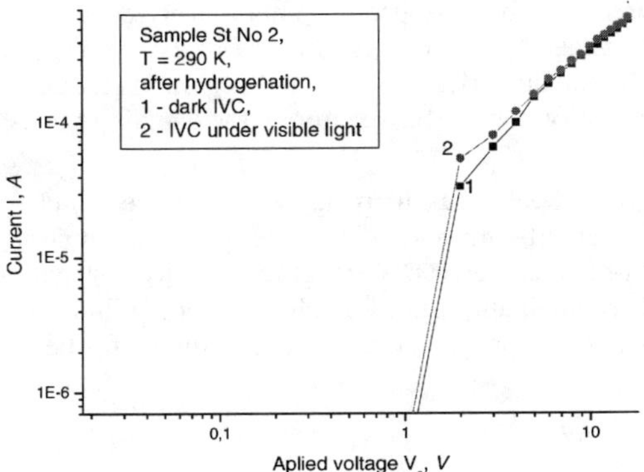

Figure 4b.

Figure 4. IVC's of the investigated samples after hydrogenation.

IVC's of the films belonging to the Set No 2 were also significantly changed (Fig. 4b). As it is shown, the first linear section due to the carriers tunneling under small bias is unchanged, unlike the second one exhibiting the velocity saturation mode of the carriers transport [6]:

$$I = T_{tun} \frac{2\varepsilon\,\varepsilon_0 V_{sat}\,A}{L^2} V_a, \quad (3)$$

where $V_{sat} = \sqrt{(2e/m^*)V_{thresh}}$ is the charge carriers saturation velocity, m* is the carriers effective mass (for our material m* ~ 0.25 m_0, V_{thresh} = 1. 22 V is the threshold voltage, under which the current modes are changed. The next important result is the appearance of the photosensitivity in the visible part of the spectrum (curve 2 in the Fig. 4b).

CONCLUSIONS

The effect of the heat treatment in the flow of molecular hydrogen on electric and photoelectric properties of thin films of narrow-gap solid solution ZnCdHgTe is investigated for the first time. It is shown: i) the films of composition corresponding to the relatively small gap (<1.0 eV, in our case the samples from the Set No 1, $E_g = 0.86\ eV$)

dramatically change the character of the current-voltage dependencies from oscillation – like onto a diode like; ii) the films of composition corresponding to the larger gap (>1.0 eV, in our case the samples from the Set No 2, $E_g = 1.13\ eV$) change not only the

character of the charge carriers transport under applied bias (from the strong tunneling to the velocity saturation mode), but also become photosensitive under the applied voltage 1.0 – 3.0 V (the range of voltage for device operation). These experimental facts may be explained qualitatively (the numerical analysis and simulation require new experiments which are in progress) as follows: 1) passivation of the dangling bonds and defects

on the surface of the film appeared as unavoidable during the sample growth; 2) possible enlargement of the carriers lifetime and diffusion length of the minority carriers due to passivation of the non-radiated defects; 3) the process of hydrogenation may sufficiently influence the surface recombination speed. Each of the possibilities proposed as the explanation of the first results presented in the paper requests further investigations.

REFERENCES

1 .G. Khlyap, P. Sydorchuk, J. Polit et al. *Mat. Res. Soc. Proc.* **Vol.785**, D10.10.1 (2004).

2. I. Martin, M.Vetter, E. Apella et al., *Appl. Phys. Lett.* **84**, 1474 (2004).
3. M. Ristova, Y. Kuo, and H.H. Lee, *Appl. Surf. Sci.* **218**, 44 (2003).
4. I. Vasilev, *Phys. Stat. Sol. (b)* **239**, 19 (2003).
5. S. Sze, *Physics of Semiconductor Devices*, Wiley, New York (1981).

6. E. Hernandez, Cryst. Res. Tech. 33, 285 (1998).

General Properties of
Hydrogen in Semiconductors

Hydrogen-Nitrogen Tailors Semiconductor Optoelectronics: The Case of Dilute Nitride III-V Alloys

A. Janotti

Metals and Ceramics Division, Center for Computational Sciences,
Oak Ridge National Laboratory, Oak Ridge, Tennessee 37831, USA

ABSTRACT

Hydrogen is an omnipresent impurity in semiconductors, often associated with other impurities and native defects, strongly affecting their electronic properties by passivating deep and shallow levels, or activating isoelectronic centers, and can be intentionally or unintentionally incorporated. On the other hand, nitrogen has profound effects on the electronic structure of conventional III-V compounds: just a few percent of N can drastically lower the band gap of GaAs making it suitable for long-wavelength optical devices; isovalent doping of GaP by N leads to a quasidirect band gap with enhanced optical functionality. The large difference in electronegativity between N and other group V elements is expected to couple with the high chemical activity of H, raising crucial questions about the behavior of H in dilute nitride alloys that theories of hydrogen in conventional semiconductors or in commom-anion nitrides are unable to answer. Here we show that N can qualitatively alter the electronic behavior of hydrogen: In GaAsN, an H atom bonds to N and can act as a donor in its own right, whereas in GaAs and GaN, H is amphoteric; Nitrogen also stabilizes the H_2^* complex, that is otherwise unstable against the formation of interstitial H_2 molecules, reversing the effect of N on the band gap of GaAs, allowing us to interpret several recent experiments.

INTRODUCTION

Adding just a few atomic percent of N in GaAs can simultaneously reduces the lattice constant *and* causes a drastic reduction of the band-gap, offering unique opportunities in band-gap engineering for laser and photovoltaic applications[1–6]. In the case of GaP, the addition of N can lead to a quasi-direct band gap with enhanced optical functionality.[7] The growth processes of these dilute nitride alloys usually involve hydrogen[8], such as in metal organic chemical vapor deposition (MOCVD) or gas-source molecular beam epitaxy (MBE)[9–11]. The large electronegativity and size mismatch between the N and other group-V anions are expected to strongly couple with the high chemical activity of H, leading to new physical and chemical effects that cannot be understood on the basis of the current theory of hydrogen in either conventional semiconductors or in common anion nitrides[12, 13].

Infra red spectroscopy measurements in N doped GaP revealed three distinct peaks at 2885.5, 2054.1, and 1049.8 cm^{-1} associated with the local vibrational modes of H-related complexes. Similar results were reported for N-doped GaAs and, to explain the data, a

H-N-H dihydride model of trigonal symmetry was proposed where two H atoms are bonded directly to a N atom[14, 15]. There are, however, several difficulties with this model: first, the N is fivefold coordinated which has never been observed in any nitride; Second, the two H stretching frequencies differ surprisingly by 830 cm^{-1}, in contrast to the ammonia molecule, where the triplet splitting is only about 100 cm^{-1}; and third, the observed isotope shifts of the two high frequency modes are distinctly different: 5.4 and 1.7 cm^{-1}. From a more fundamental point of view, diatomic H_2 complexes other than the H_2 molecules have never been predicted by theory to be stable nor have they been observed in any III-V semiconductors[16–18].

Recently, Xin *et al.* showed that hydrogen incorporation increases with nitrogen concentration in InGaAsN alloys grown by gas-source MBE. Based on Hall measurements, they suggested that H acts as an isolated donor in InGaAsN, and makes the as-grown undoped samples slightly n-type. Annealing above 700 ^{o}C reduces the hydrogen concentration and renders the samples p-type[10, 11]. This behavior contrasts with the behavior of hydrogen in other semiconductors where H acts as a passivation agent, but not as a source of doping in its own right[12]. In addition, post-growth hydrogenation of InGaAsN and GaAsN alloys can completely reverse the drastic band-gap reduction caused by N, suggesting an intriguing role of H in dilute nitride alloys[19, 20]. It was then suggested that H-induced band-gap restoration is attributed to the binding of one H atom to the N atom in a passivation process.

Using first-principles electronic structure method we investigated the structure and stability of monohydride and dihydride complexes in dilute GaPN and GaAsN alloys. We found that monatomic hydrogen is stable in the bond-center site in both p-type and n-type conditions, in contrast to the behavior of H in the parent compounds GaAs, GaP, or GaN where H is stable in the bond-center site in p-type conditions and in the Ga antibonding site in n-type conditions. In GaAsN, monatomic H can act as a donor for all Fermi level positions in the band gap. This is quite surprising because in conventional semiconductors, monatomic H can exist in either donor or acceptor charge states depending on the Fermi level position[12]. The strong chemical bond between H and N stabilizes the N-related H_2^* complexes (NH_2^*) in dilute GaPN and GaAsN alloys, that are otherwise unstable in the parent compounds GaAs, GaP, and GaN, where the interstitial H_2 molecules are the stable dihydride forms[21–23]. The calculated local vibrational frequencies and isotope shifts of the NH_2^* complex in GaPN agree with the experimental measurements[21]. In contrast, the H-N-H model where two H directly bonds to the same N atom in a trigonal complex[14, 15] is unstable against spontaneous transformation into the NH_2^*. Moreover, the formation of the NH_2^* complex in dilute GaAsN completely reverses the N-induced band-gap reduction in GaAs[22, 23]. These findings also apply to InGaAsN alloys, thus providing a qualitative explanation to the recent experimental observations[19, 20]. Similar results were then obtained by other theory groups[24–26].

THEORETICAL APPROACH

The calculations are based on the Density Functional Theory within the local density approximation (LDA)[27] in the pseudopotential framework as implemented in the VASP code[28]. The hydrogen complexes in dilute GaAsN (GaPN) alloys were simulated by adding n H atoms ($n = 1, 2$) to a 64-atom periodically repeated supercell, where one As (P) is replaced by one N atom corresponding to a N content of 3.125%. The integration of the charge density in the Brillouin zone was performed on the Monkhorst-Pack special k points, equivalent to the 4x4x4 mesh for the zincblende primitive cell. Convergence with respect to the supercell size was tested by comparing the results obtained with 32, 64, and 128 atoms supercells. We tested the convergence with respect to the number of plane waves in the basis set, and we find that at a cutoff of 23 Ry the estimated error in the total energy difference is less than 0.05 eV.

It is well known that LDA underestimates the band gap, and therefore a correction should be applied. It is also known that LDA gives reasonably accurate atomic geometries, elastic properties, and heats of formation. Thus, the LDA corrections to the defect formation energy arise mainly because of the LDA errors in the occupied single-particle eigenvalues associated with the defect[29]. One can thus project the defect state onto a set of bulk states which form a complete basis, and subsequently correct the LDA errors based on the GW quasiparticle calculations for the bulk states[30]. In this study, we derive the projections by calculating the pressure coefficient of the defect levels and decomposing it in a two-band (valence + conduction band) approximation[31].

The calculated formation energy per hydrogen that gives the stability of the hydrogen related complexes in the dilute nitride alloys is defined as,

$$E_f = \frac{1}{n}[E_{tot}(\text{host} + n\text{H}, q) - E_{tot}(\text{host}) - n\mu_\text{H} + q\varepsilon_F] \tag{1}$$

where $E_{tot}(\text{host} + n\text{H}, q)$ is the total energy of the hydrogen complex with n H atoms at the charge state q in the host, and $E_{tot}(\text{host})$ is the total energy of the pure host (either GaAsN, GaPN, GaAs or GaP). The hydrogen chemical potential μ_H is the energy per atom of the H_2 molecule in free space at T $= 0$. And ε_F is the Fermi energy measured with respect to the host valence band maximum (VBM).

DISCUSSION

Monohydride complexes in dilute GaAsN and GaPN alloys

We studied the donor (+), neutral (0), and acceptor (−) charge states of monohydride complexes in various configurations, including the bond-center next to N (NH$_{BC}$) and far away from N (H$_{BC}$), the antibonding next to N, As, and Ga (NH$_{AB}$, AsH$_{AB}$, and GaH$_{AB}$, respectively), and the tetrahedral interstitial configurations. Our results show that the NH$_{BC}$ is the lowest energy configuration in the donor, neutral, and acceptor charge states. The other configurations are at least 1 eV higher in energy except for NH$_{AB}$ (see below).

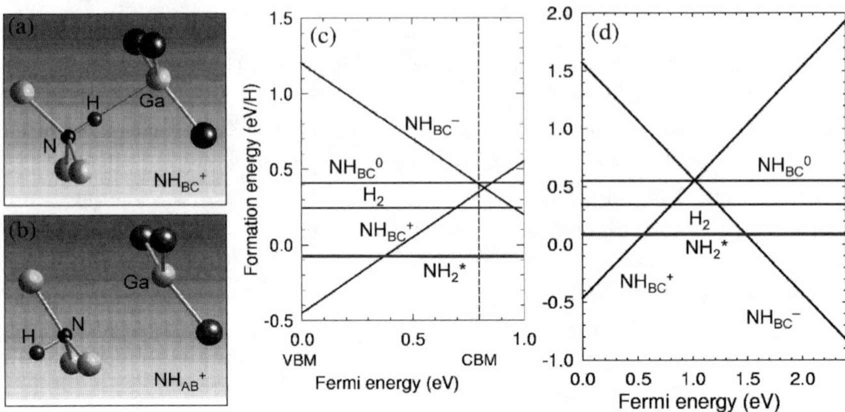

FIG. 1: Ball and stick hydrogen and model of the lowest energy monohydride complexes (a) NH$_{BC}^+$ and (b) NH$_{AB}^+$ in dilute GaAsN and GaPN alloys. The formation energy as a function of the Fermi energy ε_F of the lowest energy monohydride complexes in dilute (c) GaAsN and (d) GaPN. The formation energy of the dihydride NH$_2^*(\alpha)$ and interstitial H$_2$ molecule are also shown for comparison.

In the donor charge state NH$_{BC}^+$ [Fig. 1(a)], the H atom strongly binds to the N atom with a H-N bond length of 1.05 Å. The Ga atom is displaced along the [111] direction to the basal plane formed by its three nearest neighbor As atoms. The Ga-H distance is 50% larger than the sum of their respective covalent radii. The N atom is displaced along the [1 1 1] direction. In the donor charge state NH$_{AB}^+$ [Fig. 1(b)], the H atom at the antibonding site strongly binds to the N atom with a N-H bond length of 1.05 Å. In this case, similar to the NH$_{BC}^+$, both N and Ga are displaced away from each other to their respective basal planes along the [111] direction. These results are quantitatively similar in GaAsN and GaPN.

The NH$_{AB}^+$ is 0.4 eV higher than the NH$_{BC}^+$, mainly because the electron charge density of the host at the bond-center site is higher than that at the antibonding site. Thus, the Coulomb binding energy for H$^+$ at the bond-center site is larger. This is in contrast to pure GaN where H$^+$ prefers the N antibonding site[32]. The Ga-N bond length in the dilute GaAsN and GaPN alloys are larger, and therefore, less strained than that in GaN. The reduced strain energies are also responsible for the stabilization of the NH$_{BC}^0$ and NH$_{BC}^-$ in GaAsN and GaPN.

The formation energy as a function of the Fermi energy ε_F of the lowest energy monohydride complexes in GaAsN and GaPN are plotted in Fig. 1(c) and 1(d). In dilute GaAsN alloys, monatomic H exists predominantly in the donor charge state with the $(+/-)$ transition level above the conduction band minimum (CBM). This result is in clear contrast to those in the parent compounds GaAs and GaN, where H is an amphoteric impurity, existing as a donor in p-type conditions and as an acceptor in n-type conditions[12, 18, 32]. This difference can be explained by the exceptionally large bowing effect of N that lowers the conduction band minimum (CBM) of GaAsN below both the CBM of GaAs and GaN, as well as below the H $(+/-)$ level. In the case of dilute InGaAsN alloys, the presence of indium lowers the CBM even further below that of GaAsN. Therefore, we expect that the hydrogen $(+/-)$ level will be even further above the CBM. In general, when the host CBM level is very low monatomic H can behave exclusively as a donor[33], as has been proposed for ZnO and InN[13, 34, 35]. However, a similar behavior due solely to an alloying effect has not been suggested before.

An interesting question regarding H in p-type InGaAsN is whether it will induce compensation or passivation. Passivation occurs when an electrically neutral acceptor-donor complex is formed, while compensation implies that most donors are spatially separated from the acceptors. Compensation further implies that when hydrogen is removed by annealing, Hall mobility will increase as the number of ionized impurity scattering centers is reduced. This differs from what has normally been observed in elemental and III-V semiconductors where a decrease in the Hall mobility upon annealing suggests the removal of hydrogen from H-passivated acceptor complexes[36]. Because the H^+-N binding energy of 1.0 eV is considerably larger than typical H^+-acceptor binding energies (e.g., 0.4 eV for Be-H^+), it is, therefore, entirely possible that in dilute GaAsN and InGaAsN H compensates instead of passivating the acceptors. Indeed, in the case of Ref. 11 in which Be is used as the dopant, the measured Hall mobility shows such an anomaly and increases upon annealing.

Dihydride complexes in dilute GaAsN and GaPN alloys

We first discuss the formation of dihydride complexes in dilute GaAsN alloys. The experiments in Refs. 10 and 11 show that the total hydrogen concentration [H] is higher than the free electron concentration in as-grown unintentionally n-doped InGaAsN. This suggests that hydrogen can also be present in electrically inactive forms such as interstitial H_2 molecules or H_2^* complexes[37]. Furthermore, our calculations show that the formation of NH_{BC}^+ only weakly affects the band gap. Hence, a defect other than the active NH_{BC}^+ has to be responsible for the large band-gap opening (of several tenths of an eV) observed in the post-growth hydrogenation experiments[19, 20]. To identify this defect, we have studied different dihydride complexes that involve two hydrogen atoms adjacent to the nitrogen atom. Our results, represented in Fig. 2, show that the N induces the stabilization of H_2^*-like complexes (NH_2^*) where one H atom is bonded to the N atom and the other H atom is bonded to an adjacent Ga atom in a C_{3v} symmetry as shown in Fig. 2(a).

The $NH_2^*(\alpha)$ represented in Fig. 2(a)] is the lowest energy complex with $E_f = -0.07$ eV/H in GaAsN. One H is at the antibonding site close to N, similar to NH_{AB}^+ in Fig. 1(b), with a N-H bond length of 1.05 Å. The second H is at the bond-center site with a Ga-H bond length of 1.54 Åin GaAsN. No chemical bond is formed between this second H and N

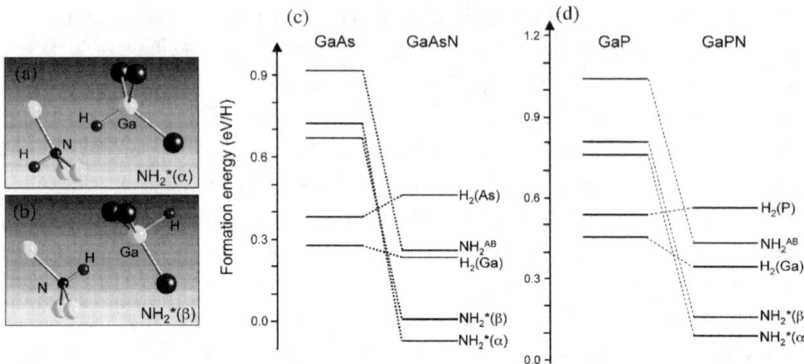

FIG. 2: Ball and stick model of the stable $NH_2^*(\alpha)$ and the metastable $NH_2^*(\beta)$ electrically inactive complexes in dilute (a) GaAsN and (b) GaPN alloys. The formation energy of dihydride complexes in (c) GaAs and GaAsN, and (d) GaP and GaPN illustrate the N-induced stabilization of H_2^* complexes in dilute nitride alloys.

as the N-H separation more than 90% larger than the sum of their covalent radii of 1.07 Å [21]. The $NH_2^*(\beta)$ represented in Fig. 2 (b) is only 0.1 eV/H higher than the $NH_2^*(\alpha)$. The N-H bond length is 1.05 Å and the Ga-H bond length is 1.60 Å. Both the N and Ga atoms assume a planar configuration, raising the strain energy of the β structure with respect to the α structure where the Ga assumes almost its original configuration in GaAs. The $NH_2^*(\alpha)$ complex is strongly favored by as much as 0.3 eV/H over the interstitial H_2 molecule in GaAsN. In contrast, in GaAs the interstitial H_2 molecule is considerably more stable than the H_2^* complexes shown in Fig. 2(c). Note that there is an "universal" reduction of 0.7 eV/H of the NH_2^* complexes in GaAsN with respect to their counterparts in GaAs. This reduction can explained by the 1.5 eV energy difference between the N-H bond and the As-H bond found in NH_3 and AsH_3 molecules.

An analysis of the density of states for GaAs, GaAsN, and $GaAsNH_2^*$, as shown in Fig. 3(a) reveals that while N incorporation lowers the CBM of GaAs by 0.6 eV for [N] of 3.125 %, the formation of $NH_2^*(\alpha)$ pushes the CBM back up completely — a robust result that holds for 32, 64, and 128 atoms supercell calculations that corresponds to N atomic concentrations of 6.25, 3.125, and 1.5625%, respectively. This remarkable effect can be qualitatively understood schematically as a two step process represented in Fig. 3(b): breaking the Ga-N bond results in a N dangling bond (DB)-like state resonant in the valence band and a Ga DB-like state near the GaAsN CBM. The binding of the first H atom to the N DB-state results in a N-H^B bonding state deep in the valence band and a N-H^A antibonding state resonant in the GaAs conduction band. The binding of the second H to the Ga DB-like state results in a bonding Ga-H^B state below the GaAs VBM and an antibonding Ga-H^A state above the

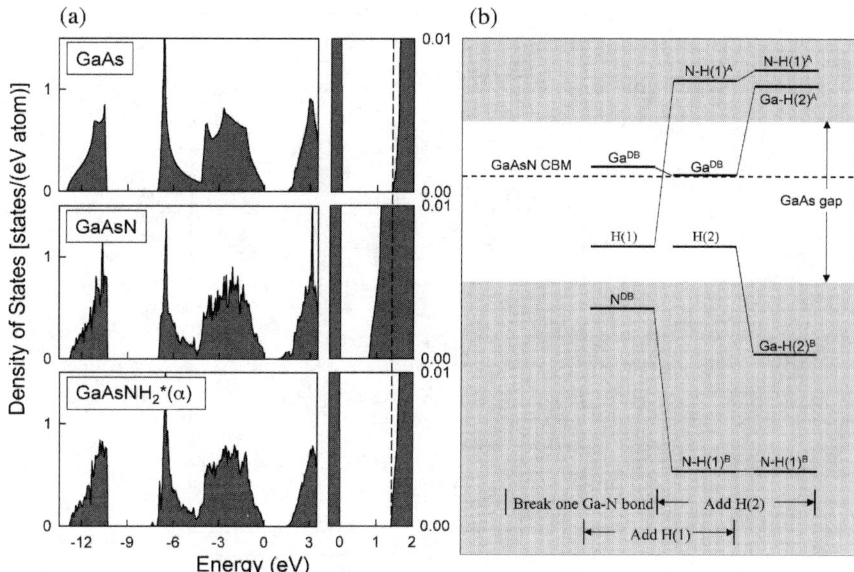

FIG. 3: (a)Total density of states for GaAs, GaAsN, and GaAsNH$_2^*$. The calculations were done for a 64-atom supercell ([N] = 3.125%). To show more clearly the effect of NH$_2^*$ on the band gap, we draw on the right panel the magnified DOS near the gap. The VBM is at the origin and the CBM of GaAs is indicated by a vertical dashed line. (b) Schematic two-step process of the N "passivation" by the NH$_2^*$(|alpha) complex (see text).

CBM of GaAs[22, 23]. The net result of this process is that one N is completely "passivated" by the formation of a NH$_2^*$ complex and does not contribute to the band gap reduction in GaAs.

The formation energy of NH$_{BC}^+$ increases as the Fermi energy ε_F increases (Fig. 1(c) and 1(d)). In Fig. 4 we show the results of a detailed balance analysis between NH$_{BC}^+$ and NH$_2^*$. Assuming that the [H] is less than [N], the detailed balance analysis shows that the relative concentration [NH$^+$]/[H] decreases with [H] whereas [NH$_2^*$]/[H] increases. Interestingly, however, the absolute [NH$^+$] increases with [H] instead of decreasing. As a result, the Fermi energy also increases with [H]. At [H] = 10^{19} cm^{-3}, the calculated [NH$^+$] of 10^{16} cm^{-3} agrees with the experimental free-electron concentration of 7×10^{15} cm^{-3}[11]. This suggests that the majority of the single-H are ionized so the (+/−) transition level is indeed shallow. When [H] is comparable to [N], however, the role of H is shifted towards restoring the GaAs band gap by the formation NH$_2^*$.

As H is gradually introduced during post-growth hydrogenation as reported in Refs. 19, 20 the concentration of NH$_2^*$ increases whereas the concentration of non-hydrogenated N

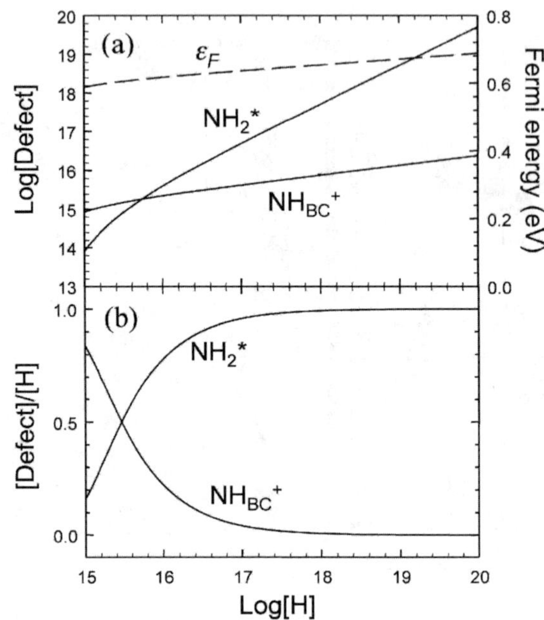

FIG. 4: Detailed balance analysis of the formation of the NH_2^* versus the NH_{BC}^+ in dilute GaAsN alloy at T=350 K. (a) The NH_{BC}^+ and NH_2^* concentration (left scale), and the Fermi energy measured from the VBM (dashed line and right scale) as a function of the total H concentration [H]. (b) The relative concentration of NH_{BC}^+ and NH_2^* as a function of [H]. It is important to note that [H] is assumed to be less than [N] so that the H effect on the band gap is negligible. Results calculated at other temperatures are qualitatively the same.

decreases. The band gap also increases because the N-induced gap reduction is proportional to the concentration of non-hydrogenated N. This process approaches completion when [H] is more than two times larger than [N], which leads to the exposure of the original GaAs band gap. The same conclusion can also be straightforwardly applied to InGaAsN alloys, which have an even lower CBM than GaAsN.

As in GaAsN, N plays a pivotal role in stabilizing NH_2^* complexes over the H_2 molecules. The $NH_2^*(\alpha)$ is also the lowest energy complex with $E_f = 0.09$ eV/H in GaPN . The $NH_2^*(\beta)$ is 0.1 eV/H higher than the $NH_2^*(\alpha)$. In the case of the $NH_2^*(\alpha)$, the antibonding H-N bond length is 1.05 Å, and the bond-center H-Ga bond length is 1.53 Å. The distance between the bond-center H and N is more than 80% larger than the sum of their atomic radii of 1.07 and, therefore, no chemical bond forms between the two atoms as can be seen in the total charge density plot[21]. In contrast, the H-N-H model proposed in Refs. 14, 15 is found to be unstable: It spontaneously transforms into the $NH_2^*(\alpha)$ complex without any energy barrier. We also studied a NH_2^* configuration where both H are at the antibonding positions, one close to N and the other close to Ga (NH_2^{AB}). In this case, the H-N and H-Ga bond lengths are 1.05 and 1.61 Å, respectively. The Ga and N atoms are displaced along the [111] and [$\bar{1}\,\bar{1}\,\bar{1}$] directions, respectively, to form planar configurations with their nearest neighbor host atoms. This complex is 0.3 eV/H higher than the $NH_2^*(\alpha)$ and, therefore, it unlikely forms.

Fig. 2(c) and 2(d) show how N affects the energy of the various H pairs in GaAsN and GaPN. First, regarding the interstitial H_2 molecules, nitrogen slightly raises the energy of $H_2(V)$, but lowers the energy of $H_2(Ga)$ by 0.1 eV/H, where $H_2(V)$ is located in the tetrahedral interstitial next to the group V element and $H_2(Ga)$ is located in the tetrahedral interstitial next to the Ga. The reason is that N is more electronegative than P so it takes more electrons *away* from the vicinity of the Ga than P does. Because interstitial H_2 prefers low charge density regions, the net result of the charge transfer is that $H_2(Ga)$ is even more stable than $H_2(N)$. Second, N systematically reduces the energies of the NH_2^* complexes by about 0.7 eV/H, as in GaAsN. This is explained by the 1.5 eV energy difference between the N-H and P-H bonds in the NH_3 and PH_3 molecules.

The unusual strong H-N bond raises the question whether the H pairs are stable against dissociation into two individual H-N complexes, especially when the N concentration is larger than H. This would maximize the number of H-N bonds. We have calculated the $NH_2^*(\alpha)$ binding energy with respect to distant H-N pairs of various charge states. We find that for $\varepsilon_F \leq 0.5$ eV, the formation of two separated NH_{BC}^+ is energetically favorable with respect to the formation of one $NH_2^*(\alpha)$. For $0.5 \leq \varepsilon_F \leq 1.5$ eV the formation of $NH_2^*(\alpha)$ is energetically favorable. The experiments in Refs. 14, 15 were performed with intrinsic N-doped GaP and GaAs where the Fermi energy ε_F is near the middle of the band gap. Hence, H_2^* is expected to be the most abundant among the H centers.

We investigate the local vibration modes of NH_2^* in GaPN. The vibrational frequencies are calculated by evaluating the force-constant matrix K where the elements K_{ij} are given by the Hellman-Feynman force induced on the ith atom by the displacement of the jth atom[38]. We first consider only the harmonic approximation. Anharmonic effects are removed by averaging the forces for both positive and negative displacements. The results are shown in Table I.

We note that: (1) The highest frequency Mode-1 at 3081.2 cm^{-1} is associated with the stretching of the H-N bond. The calculated isotopic shift from ^{14}N to ^{15}N is 6.3 cm^{-1}. Both are in reasonable agreement with experimental values of 2885.5 and 5.8 cm^{-1}, respectively[14, 15]. As expected from the NH_2^* model, this H-N mode is little affected by the ^{69}Ga-^{71}Ga isotopes. (2) The middle-range frequency Mode-2 at 2051.3 cm^{-1} can be associated with the bond-center H-Ga bond stretching. It changes by 0.4 cm^{-1} if one replaces both ^{14}N by ^{15}N and ^{69}Ga by ^{71}Ga. Experimentally, the isotopic shift for this mode is 1.7 cm^{-1}. Because the bond-center H is bonded to the heavier Ga atom, a qualitatively smaller isotope shift

TABLE I: Local vibrational frequencies and isotope shifts (in cm^{-1}) for the NH$_2^*$ complexes in GaPN. Mode-1 and 2 are the stretching modes whereas Mode-3 is a doubly degenerate wagging mode. The calculated results are compared to experiment[14, 15]. Also see Ref. 39 for the effect of anharmonicity on the vibration frequencies. The anharmonic effect for Mode-1 of α-H$_2^*$ shifts the frequency downward by 167.7 cm^{-1}to 2913.5 cm^{-1}.

Mode		1 (shift)	2 (shift)	3 (shift)
NH$_2^*(\alpha)$	Calc.	3081.2 (6.3)	2052.3 (0.4)	968.3 (1.2)
	Expt.	2885.5 (5.4)	2054.1 (1.7)	1049.8
NH$_2^*(\beta)$	Calc.	3146.8	1682.9	852.1

for Mode-2 is expected. The calculated frequency difference of Mode-1 and Mode-2 is 1028 cm^{-1}, or 861 cm^{-1} after including anharmonic effects[39]. This is in good agreement with the experimental value of 831 cm^{-1}. This frequency difference cannot be explained by the previously proposed H-N-H dihydride model[14, 15], because one would expect a much smaller frequency difference if both H were bonded to N. Furthermore, the reduction of the H-N frequency (2885.5 cm^{-1}) in the NH$_2^*(\alpha)$ relative to the frequency of the normal mode in ammonia (3444 cm^{-1}) can be explained by the host charge induced bond weakening in solids[40]. A similar reduction is also expected if NH$_2^*$ is negatively charged[15]. (3) Finally, the lowest frequency mode was found to be doubly degenerate at 968.3 cm^{-1} and it is related to the wagging of the H-N bond. The isotope shift of 1.2 cm^{-1} in this case is caused by replacing ^{14}N by ^{15}N.

SUMMARY

In summary, we investigated the effects of N on the hydrogen behavior in dilute GaAsN and GaPN allooys. We find that N qualitatively alters the electronic behavior of monatomic hydrogen in GaAsN, making it *only* as a donor, despite the fact that in GaAs or GaN hydrogen is amphoteric, existing in either in donor or acceptor states. We also found that N induces the stabilization of NH$_2^*$ complexes that are otherwise unstable in the parent compounds GaAs, GaP, or GaN. The formation of NH$_2^*(\alpha)$ complex completely "passivates" N in dilute GaAsN alloys, eliminating the N-induced band-gap reduction in GaAs, providing a quantitative explanation to the recent puzzling experimental observations. The calculated local vibration modes and the isotope shifts of NH$_2^*(\alpha)$ complex can also explain the infra red spectroscopy measurements of H-related peaks in N-doped GaP and GaAs, In contrast, the previously proposed trigonal H-N-H dihydride model for H complexes is found to be unstable against spontaneous transformation into NH$_2^*$ complexes.

ACKNOWLEDGMENTS

The author acknowledges the valuable collaboration with Dr. Su-Huai Wei and Dr. S. B. Zhang form the National Renewable Energy Laboratory (CO) and Dr. Chris Van de Walle from Parc Research Center (CA). This document describes activities performed under contract number DE-AC0500OR22750 between the U.S. Department of Energy and Oak Ridge Associated Universities.

REFERENCES

[1] M. Weyers, M. Sato, and H. Ando, Jpn. J. Appl. Phys. (Part 1) **31**, L853 (1992).

[2] J. Neugebauer and C. G. Van de Walle, Phys. Rev. B **51**, 10568 (1995).

[3] S.-H. Wei and A. Zunger, Phys. Rev. Lett. **76**, 664 (1996).

[4] M. Kondow, K. Uomi, A. Niwa, T. Kitatani, S. Watahiki, and Y. Yazawa, Jpn. J. Appl. Phys. (Part 1) **35**, 1273 (1996).

[5] S. R. Kurtz, A. A. Allerman, E. D. Jones, J. M. Gee, J. J. Banas, and B. E. Hammons, Appl. Phys. Lett. **74**, 729 (1999).

[6] D. J. Friedman, J. F. Geisz, S. R. Kurtz, and J. M. Olson, J. Cryst. Growth **195**, 409 (1998).

[7] M. Singh and J. Weber, Appl. Phys. Lett. **54**, 424 (1989).

[8] M. Sato, in *GaN and related Materials*, edited by Stephen J. Pearton (Gordon and Breach Science Publishers, The Netherlands, 1997), Vol. 32.

[9] S. Kurtz, J. Webb, L. Gedvilas, D. Friedman, J. Geisz, J. Olson, R. King, D. Joslin, and N. Karam, Appl. Phys. Lett. **78**, 748 (2001).

[10] H. P. Xin, C. W. Tu, and M. Geva, Appl. Phys. Lett. **75**, 1416 (1999).

[11] H. P. Xin, C. W. Tu, and M. Geva, J. Vac. Sci. Technol. B **18**, 1476 (2000).

[12] S. K. Estreicher, Mat. Sci. Engr. Reports **14**, 319 (1995).

[13] S. Limpijumnong and C. G. Van de Walle, Phys. Stat. Sol. (b) **228**, 303 (2001).

[14] B. Clerjaud, D. Cote, W.-S. Hahn, A. Lebkiri, W. Ulrici, and D. Wasik, Phys. Rev. Lett. **77**, 4930 (1996).

[15] B. Clerjaud, D. Cote, W.-S. Hahn, A. Lebkiri, W. Ulrici, and D. Wasik, Phys. Stat. Sol. (a) **159**, 121 (1997).

[16] K. Murakami, N. Fukata, S. Sasaki, K. Ishioka, M. Kitajima, S. Fujimura, and J. Kikuchi, and H. Haneda, Phys. Rev. Lett. **77**, 3161 (1996).

[17] J. Vetterhöffer, J. Wagner, and J. Weber, Phys. Rev. Lett. **77**, 5409 (1996).

[18] L. Pavesi and P. Giannozzi, Phys. Rev. B **46**, 4621 (1992).

[19] G. Baldassarri H. v., H. M. Bissiri, A. Polimeni, M. Capizzi, M. Fischer, M. Reinhardt, and A. Forchel, Appl. Phys. Lett. **78**, 3472 (2001).

[20] A. Polimeni, G. Baldassarri H. v., H. M. Bissiri, M. Capizzi, M. Fischer, M. Reinhardt, and A. Forchel, Phys. Rev. B **63**, 201304(R) (2001).

[21] A. Janotti, S. B. Zhang, and Su-Huai Wei, Phys. Rev. Lett. **88**, 125506 (2002).

[22] A. Janotti, S. B. Zhang, Su-Huai Wei, and C. G. van de Walle, Phys. Rev. Lett. **89**, 86403 (2002).

[23] A. Janotti, S. B. Zhang, Su-Huai Wei, and C. G. van de Walle, Optical Materials **25**, 261 (2004).

[24] Yong-Sung Kim and K. J. Chang, Phys. Rev. B **66**, 155402 (2002).

[25] W. Orellana and A. C. Ferraz, Appl. Phys. Lett. **81**, 481 (2002).

[26] A. Amore Bonapasta, F. Filippone, P. Giannozzi, M. Capizzi, and A. Polimeni, Phys. Rev. Lett. **89**, 216401 (2002).

[27] P. Hohenberg and W. Kohn, Phys. Rev. **136**, B864 (1964). W. Kohn and L. J. Sham, Phys. Rev. **140**, A1133 (1965).

[28] G. Kresse and J. Furthmüller, Phys. Rev. B. **54**, 11169 (1996); G. Kresse and J. Furthmüller, Comput. Mat. Sci. **6**, 15 (1996).

[29] Q.-M. Zhang and J. Bernholc, Phys. Rev. B. **47**, 1667 (1993).

[30] M. S. Hybertsen and S. G. Louie, Phys. Rev. B **34**, 5390 (1986); S. B. Zhang, et al., Phys. Rev. B **40**, 3162 (1989); E. L. Shirley, et al. Phys. Rev. Lett. **69**, 2955 (1992).

[31] A. Janotti, S. B. Zhang, S.-H. Wei, and C. G. Van de Walle, unpublished.

[32] J. Neugebauer and C. G. Van de Walle, Phys. Rev. Lett. **75**, 4452 (1995).

[33] S.-H. Wei and A. Zunger, Appl. Phys. Lett. **72**, 2011 (1998).

[34] C. G. Van de Walle, Phys. Rev. Lett. **85**, 1012 (2000).

[35] C. G. Van de Walle and J. Neugebauer, Nature **423**, 626 (2003).

[36] N. M. Johnson, in J. I. Pankove and N. M. Johnson, Eds., *Hydrogen in Semiconductors, Semiconductors and Semimetals*, Vol. 34, (Academic Press, Boston, 1991), p. 113; J. Chevallier, B. Clerjaud, and B. Pajot, *ibid.*, p. 447.

[37] K. J. Chang and D. J. Chadi, Phys. Rev. Lett. **62**, 937 (1989); Phys. Rev. B **42**, 7651 (1990).

[38] John E. Northrup, Phys. Rev. B. **39**, 1434 (1989).

[39] The anharmonic effect (AE) is calculated following the procedure proposed by Chris G. Van de Walle (Ref. 40). For α-H_2^*, the AE lowers the frequency by 168 and 42 cm^{-1} for Mode-1 and Mode-2, respectively, but increases the frequency by 2 cm^{-1} for Mode-3. As expected, the AE for Mode-1 is significantly larger than for Mode-2 and -3.

[40] Chris G. Van de Walle, Phys. Rev. Lett. **80**, 2177 (1998).

Determination of Hydrogen in Semiconductors and Related Materials by Cold Neutron Prompt Gamma-ray Activation Analysis

Rick L. Paul
Analytical Chemistry Division
National Institute of Standards and Technology
Gaithersburg, MD 20899

ABSTRACT

An instrument for prompt gamma-ray activation analysis (PGAA) at the NIST Center for Neutron Research has proven useful for the measurement of hydrogen and other elements in a variety of materials. The sample is irradiated by a beam of low energy neutrons. Gamma-rays emitted by atomic nuclei upon neutron capture are measured and elemental concentrations determined by comparison with appropriate standards. The detection limit for hydrogen is < 5 mg/kg in most materials, and 2 mg/kg for hydrogen measured in silicon. The instrument has been used to measure hydrogen mass fractions of < 100 mg/kg in high purity germanium, and < 10 mg/kg in quartz. More recently PGAA has been used to measure hydrogen in 1 μm thick porous thin films on a silicon substrate, and in crystals of silicon carbide and cerium aluminate.

INTRODUCTION

The presence of hydrogen in semiconductor materials is known to alter electrical properties. However, quantitation of hydrogen at the levels present in these materials (mg/kg levels or lower) is difficult by most analytical methods. We have used cold neutron prompt gamma-ray activation analysis to measure hydrogen in a wide variety of materials, including semiconductors. The sample is irradiated by a beam of low energy neutrons; gamma-rays emitted by atomic nuclei upon neutron capture are measured using a high purity germanium detector. The analysis is both multielement and nondestructive, and the analyte is measured in situ. Furthermore, because both neutrons and gamma-rays penetrate the sample, the entire sample is analyzed. The presence of hydrogen is indicated by a 2223 keV gamma ray.

EXPERIMENT

The cold neutron PGAA spectrometer (Fig. 1) is located in the cold neutron guide hall of the NIST Center for Neutron Research (NCNR). This instrument has been described previously.[1-3] Neutrons from the reactor core, moderated by passage through liquid hydrogen at 20 K, pass through a ^{58}Ni coated guide to the sample position of the PGAA station. The neutron beam is collimated to a diameter of 2 cm or smaller before striking the sample. The neutron fluence rate at the sample position is 1×10^9 cm^{-2} s^{-1}.

Samples for irradiation are normally sealed into bags of FEP Teflon and mounted between Teflon strings suspended between the prongs of an aluminum fork. The optimum sample size is normally between 0.1 g and 1 g, although samples smaller or larger than this may be analyzed. Samples may be irradiated in air, or inside an evacuated magnesium sample chamber that reduces the hydrogen background by a factor of 2.5. Due to the presence of another neutron guide, 3 cm above the center of the PGAA beam, the size of the sample that can be mounted is limited. Furthermore, the interior areas of samples larger than about 6 cm diameter cannot presently be analyzed.

Gamma rays emitted upon neutron capture are measured by a cadmium and lead shielded high resolution germanium detector. In order to minimize the spectrum baseline continuum that results from Compton scattering, the germanium detector is surrounded by a bismuth germanate (BGO) Compton shield. Gamma-ray signals are processed using appropriate electronics modules, and gamma-ray spectra up to 10 MeV are collected and analyzed using a computer workstation.

Because of hydrogen free construction, compact geometry, low background from reactor gamma-rays, and higher capture rates for cold neutrons (vs. thermal), the limit of detection for hydrogen is better than that obtained by most thermal neutron PGAA instruments, and is comparable to or better than that obtained by other facilities with guided beams and cold neutrons. Only a few such facilities exist worldwide.[2] The detection limit for hydrogen is < 5 mg/kg in most materials, and is 2 mg/kg for hydrogen measured in silicon.

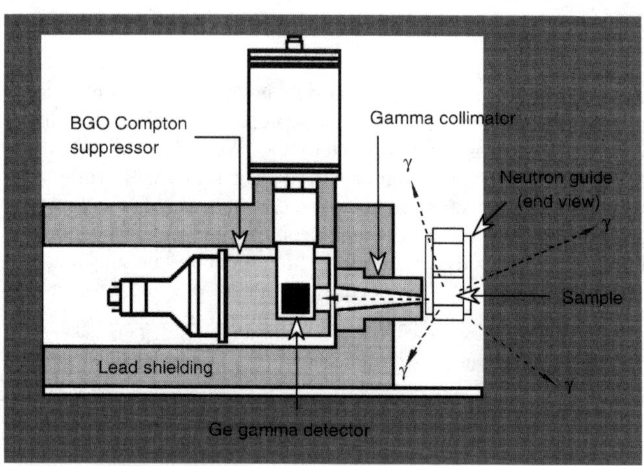

Figure 1. Side view of the cold neutron prompt gamma-ray activation analysis spectrometer in the NIST Cold Neutron Research Facility.

RESULTS AND DISCUSSION

We have previously measured hydrogen mass fractions of 50 ± 40 mg/kg to 80 ± 55 mg/kg (2s uncertainty based on counting statistics) (corresponding to 0.4 to 0.6 atomic percent of hydrogen) in five crystals of semiconductor grade germanium.[4] Samples which had been electrolytically etched prior to analysis were compared with unetched samples in order to determine whether etching of samples results in greater uptake of hydrogen. No significant differences in hydrogen concentration were observed between the etched and unetched crystals. In another study, hydrogen mass fractions of < 10 mg/kg were measured in crystals of hydrothermally grown quartz from the U. S. Army Research Laboratory (ARL).[4] The latter results were in agreement with hydrogen values determined from infrared spectroscopy measurements at ARL.

More recently, hydrogen was measured in 1 μm thick porous thin films of silicon on a single crystal silicon substrate. Porous thin films are slated to be the next generation of thin films used in integrated circuits . The atomic content of hydrogen in these films had been measured previously by RBS (Rutherford backscattering) and FRES (forward recoil elastic spectrometry). The PGAA measurements provided an independent measurement of the hydrogen content and were valuable to the research effort. The results of the measurements are given in Table 1. PGAA results are compared with hydrogen measured by two other methods. The first method uses x-ray reflectivity to measure the thickness and electron density. The atomic density is then calculated using atomic ratios of H, C, O, and Si from RBS and FRES. The second method uses FRES alone for the atomic composition. It should be noted that FRES and RBS measure only surface hydrogen, while PGAA measures total hydrogen (film + substrate). This may explain why the PGAA results are higher than the results of the other methods.

Table 1 – Hydrogen mass fractions measured in porous films on a silicon substrate. The H/Si ratio film + substrate was measured by PGAA. Uncertainties are 2s, based on counting statistics.

Sample	mg H/kg Si (PGAA)	mg H/kg Si (XR + FRES + RBS)	mg H/kg Si (FRES)
Si 26-1	26 ± 4	13.7	14.9
Si 26-2	22 ± 4		
Si 27-1	11 ± 6	7.8	6.3
Si 27-2	11 ± 3		
Si 28-1	28 ± 5	9.0	12.5
Si 28-2	23 ± 7		

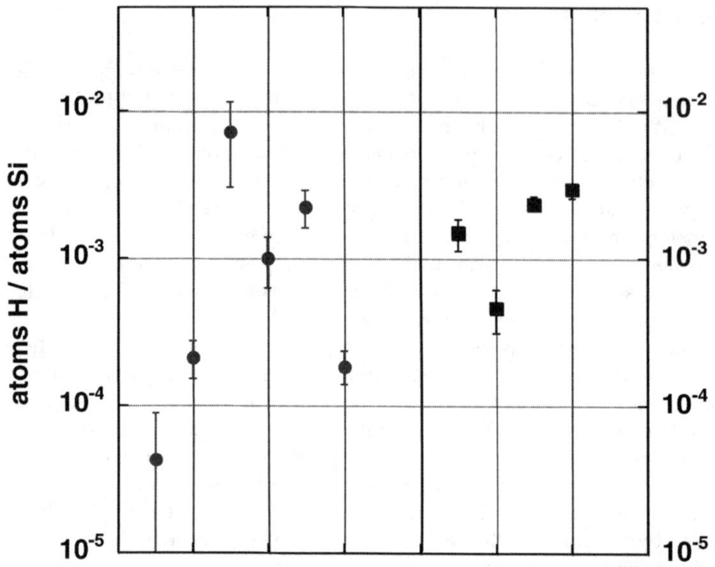

Figure 2. Hydrogen fractions measured in samples of silicon carbide (●) and cerium aluminate (■). Uncertainties are 2s based on counting statistics, background subtraction, and peak fitting.

PGAA has also been used to measure hydrogen in crystals of silicon carbide and cerium aluminate in an effort to determine the quality of these crystals. The results of those measurements are shown in Figure 2. The hydrogen concentrations of these materials were found to span several orders of magnitude. Even at low H concentrations (atom fractions of 10^{-4} to 10^{-3}), it is possible to distinguish between samples whose H content varies by a factor of 2.

CONCLUSIONS AND FUTURE STUDIES

Hydrogen mass fractions of less than 10 mg/kg have been measured in semiconductors and other materials. In the future, the applicability of the technique may be enhanced by applying existing technologies and planned upgrades to the instrument. The application of a neutron lens[5,6] , that can focus the beam down to a spot size of 0.5 mm with minimum loss of flux, has made it possible to probe the elemental content of materials as a function of position. Future plans to bend the PGAA neutron beam away from the upper portion of neutron guide NG7 will permit mounting of larger samples in the beam. This will make it possible to analyze the elemental contents of larger samples (e.g. a 200 mm silicon wafer) as a function of position.

Acknowledgments

The author would like to thank Dr. Richard Lindstrom for his advice and assistance, and the following people for providing samples for analysis: Dr. Samuel Trevino of the U.S. Army Research Laboratory, the late Dr. Richard Deslattes formerly of the NIST Physics Laboratory, Dr. Barry Bauer of the NIST polymers division, and Dr. Andreas Magerl and Ms. Elena Machkina from the University of Erlangen-Nürnberg. The author would also like to thank the staff of the NIST Center for Neutron Research.

References

1. R. L. Paul, R. M. Lindstrom, and A. E. Heald, *J. Radioanal. Nucl. Chem.*, **215 (1)** 63-68. (1997)
2. R. L. Paul and R. M. Lindstrom, *J. Radioanal. Nucl. Chem*, **243**, 181 – 189, (2000).
3. R. L. Paul, *American Laboratory*, **34(3)**, 15–20 (2002).
4. R. L. Paul and R. M. Lindstrom, in *Diagnostic Techniques for Semiconductor Processing*, edited by O. J. Glembocki, S. W. Pang, F. H. Pollak, G. M. Crean, G. Larrabee, (Mater. Res. Soc. Proc. **324**, Pittsburgh, PA 1994), pp. 403-408.
5. H. H. Chen-Mayer, D. F. R. Mildner, V. A. Sharov, Q. R. Xiao, Y. T. Cheng, R. M. Lindstrom, and R. L. Paul, 68(10) (1997) 3744 – 3750.
6. H. H Chen-Mayer, E. A. Mackey, R. L. Paul, D. Mildner, *J. Radioanal. Nucl. Chem*, **244 (2)**, 391 – 397, (2000).

Hydrogen induced degradation in GaInP/GaAs HBTs revealed by low frequency noise measurements

J.G. Tartarin[1], L. Escotte[1], M. Borgarino[2], R. Plana[1], J. Graffeuil[1]
[1] Paul Sabatier University and LAAS-CNRS, 07 av. Col. Roche, 31077 Toulouse cedex 4, France.
[2] University of Parme, Italy

ABSTRACT

One of today's challenges to enable the improved electrical performances and reliability of microelectronic devices consists in controlling impurities contamination: hydrogen appears to be present in most (if not all) the processes steps of the devices making (ambient atmosphere, or associated with AsH_3 -VPE or $AsCl_3$ -VPE for example in GaAs based devices,…). Hydrogen induced reliability has already been investigated for many Si or GaAs based technologies ((C)MOS, FET, HEMT, PHEMT as well as HBT devices). These effects of hydrogen on electrical behavior and on long term reliability are very difficult to understand because of the different nature and ionic association of hydrogen (H, H_+, H., H_2, or associated with impurities (Ge-H, Be-H, C-H,…). Most of these studies make use of IR, SIMS, Hall measurements: in this paper, we use low frequency noise measurements, associated with static as well as dynamic characterization to identify the degradation process in GaInP/GaAs Heterojunction Bipolar Transistors (HBT supplied by Thomson LCR). The presence of Hydrogen has been identified by DLTS at the LPSC laboratory (Meudon, France). The influence of passivation (SiN and GaInP ledge) on the reliability associated with Hydrogen has been one of the first improvements on HBT devices. Low-frequency noise measurements have been performed in the range of 250Hz to 100 kHz. The noise spectra evolutions (current and voltage noise sources at the input of the devices) allowed us to identify the activation process responsible of the static and dynamic rise and fall of the HBT's current gain. Chemical reactions of C-H complexes have been proved to be the processes responsible of this degradation. Additive reliability tests have been performed on two sets of devices (featuring different emitter length) under two distinct stocking conditions (temperature and biasing of the devices) leading to different junctions temperatures: low-frequency (LF) noise measurements, associated with static and dynamic S parameter measurements led to the same conclusion about the involved chemical reaction. We found that C-H complexes break, and H diffusion towards the extrinsic surface of the device has been observed on the measured leakage currents. Sealed devices have proved to get the same degradation signature than on wafer devices: Hydrogen is assumed to be present in high concentration levels in the device layers, and reacts under thermal and electrical stress.

INTRODUCTION

Some papers [1][2] have evidenced the influence of the passivation layer on the common emitter current gain β degradation of HBT devices: GaInP passivation ledge have been found to procure a good stability of the current gain $\beta=I_C/I_B$ during the stress time (compared to SiN passivation layer) and this technique is largely used by HBT manufacturers (Thomson [3] and Fujitsu [4] for example) . In this work, the devices have been processed with this GaInP passivation ledge: β fluctuations are already under the scope of these devices reliability, and thermal / electrical stress have been applied to the HBTs in order to give evidence of the failure mode mechanism involved, leading to the electrical performances degradation process. These degradation mechanisms have been proved to be strongly correlated to surface states [1][2].

GaInP/GaAs HBTs have been processed at LCR-Thomson research center, using a LP-MOCVD growing technique. The base layer (120nm thick) is carbon doped ($N_A=5.10^{19}cm^{-3}$). Non self aligned technology has been employed for the making of the HBT devices featuring $2x40\mu m^2$ emitter area (#A lot: 5 devices) and $2x30\mu m^2$ emitter area (#B lot: 5 devices). The devices are submitted to thermal and DC stress conditions: these devices are biased under a collector-emitter voltage $V_{CE}=9V$, and collector current densities respectively of $J_C=12.5$ kA/cm² and $J_C=16.7$ kA/cm² at ambient temperatures respectively of Ta=40°C and Ta=70°C (i.e. junction temperatures of 73°C for #A lot, resp. 115°C for #B lot). The HBTs are packaged into BMH60 fixtures and the cover is stick with conductive H20E glue. The two lots are submitted to two different accelerating stress factors in order to conclude to the identical failure signature (only the degradation's speed changes) for all the devices independently from the collector current density or temperature levels. Devices from #A and #B are exposed to different junction temperatures, low enough to avoid the appearance of other failure mechanism.

In the first paragraph, we give some statistical behaviors of the two distinct lots, for electrical parameters issued from DC measurements and for current and voltage noise densities issued from LF noise measurements. The second paragraph presents the experimental stress results, through the identification of the passivation of carbon by hydrogen in heavily C-doped GaAs epitaxial layer.

STATIC AND LOW-FREQUENCY NOISE CHARACTERIZATION BEFORE STRESS

Measurements have been performed before and after each stress period up to 470 hours for #B lot (final stress time for #A lot : 740 hours). Static and low-frequency noise measurements are plotted versus electrical parameters. From figure 1 and figure 2, we can appreciate the statistical scattering on the two lots, and the relation linking the noise sources with electrical (physical) parameters.

Static measurements are performed with an HP4142 system, and LF noise measurements are achieved with an experimental setup based on the multi-impedance method.

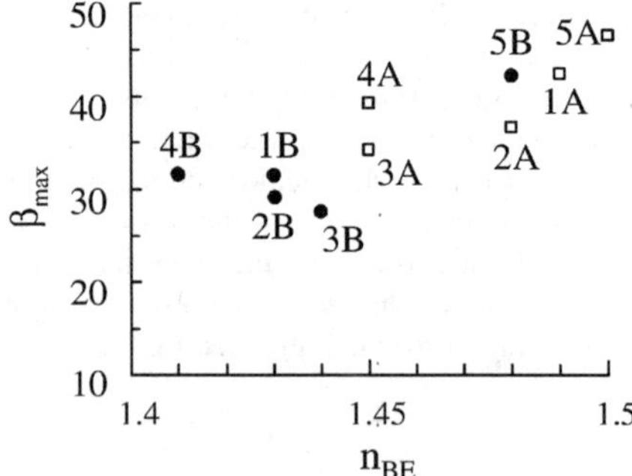

Figure 1: maximum common emitter current gain β_{max} versus the base-emitter ideality factor n_{BE} before any stress, for the devices from #A lot (emitter area $2x40\mu m^2$) and #B lot ($2x30\mu m^2$). The devices are biased at $V_{CE}=3V$.

Up to 18 different parameters (I(V) output characteristics, emitter, collector and base resistances, junction ideality coefficients n_{BE} and n_{BC}, leakage currents (base, collector), … are issued from DC measurements. The common emitter current gain β_{max} and the two LF noise spectral densities (current noise source S_I and voltage noise source S_V at the input of the device, measured at a frequency of 1 KHz) are correlated to n_{BE} and R_E in figure 1 and figure 2.

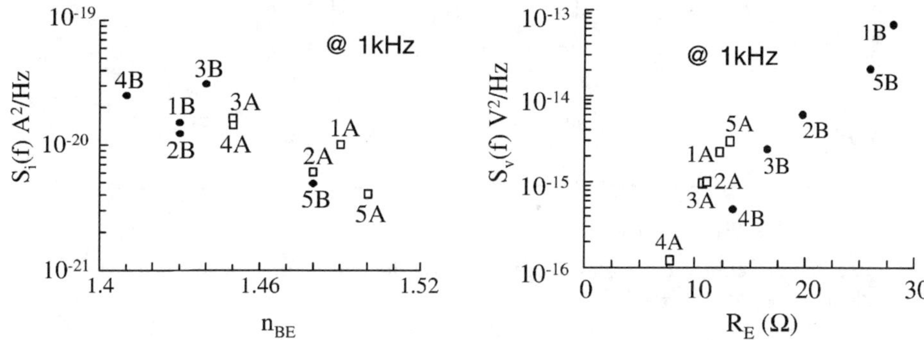

Figure 2: LF noise spectral densities S_I and S_V versus respectively n_{BE} and R_E before any stress, for the devices from #A lot (emitter area $2 \times 40 \mu m^2$) and #B lot ($2 \times 30 \mu m^2$). The devices are biased at $V_{CE}=2V$ and $I_C=3mA$.

The increase of β_{max} with n_{BE} indicates that the leakage current on the base decreases when n_{BE} increases. We also notice on figure 2 that the S_I current noise source (flicker noise) is closely correlated to surface states at the emitter-base (EB) junction. According to the simplified expression (1), this stipulates that the base surface current is lower for high n_{BE} devices (higher β_{max}, lower S_I).

$$I_B = I_{R\text{-}SCR} + I_{R\text{-}S} + I_{R\text{-}B} \tag{1}$$

I_B is the total base current composed by $I_{R\text{-}SCR}$ (space charge region recombination associated to $n_{BE}=2$), $I_{R\text{-}S}$ (surface recombination, $n_{BE}=1$) and $I_{R\text{-}B}$ (bulk recombination, $n_{BE}=1$).

The voltage LF noise source S_V is reported versus the emitter resistance: the linear dependence is related to excess noise in the emitter zone.

STATIC AND LOW FREQUENCY NOISE PERFORMANCES VERSUS STRESS TIME

The behavior of the current gain β_{max} versus the stress time features a classical shape (figure 3), with a rapid burn-in increase in the first hours of stress, and then a decrease. The same behavior is found for #A and #B lots, but the higher junction temperature for devices from #B provides earlier β degradation. The β_{max} increase is attributed to a reduction of the base surface recombination current $I_{R\text{-}SCR}$ as noticed on the evolution of n_{BE} junction ideality coefficient issued from the Gummel-plots measurements (figure 4): we find that n_{BE} increases during the stress (quasi-linear dependence with time), from about 1.4 before applying the stress to more than 2.2 at the end of the study, in accordance with $I_{R\text{-}S}$ diminution (improved surface state). Moreover, the decrease of β_{max} is associated to $I_{R\text{-}SCR}$ increase (figure 4), and not to $I_{bl,e}$ [5]

leakage current increase at the base-emitter peripheral (figure 3, right): from electrical calculations, we find the $I_{bl,e}$ contribution negligible in the overall base current.

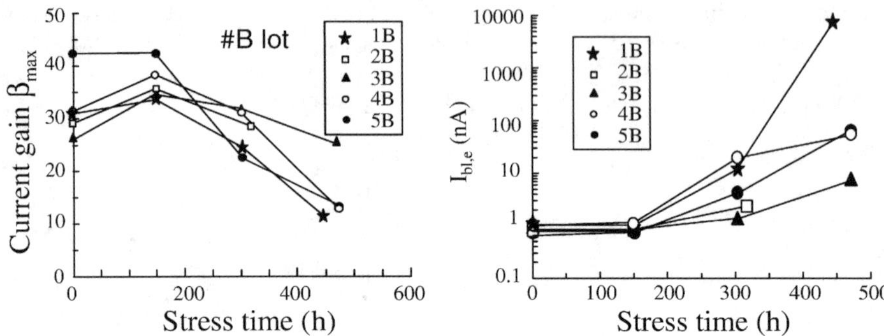

Figure 3: β current gain and $I_{bl,e}$ leakage current at the B-E peripheral versus the stress time (#B lot, devices biased at V_{CE}=3V)

In order to understand the probable chemical mechanism involved, leading to the electrical parameters alteration, we make use of LF noise measurements. Figure 5 and figure 6 show the evolution of S_I and S_V sources during the stress time.

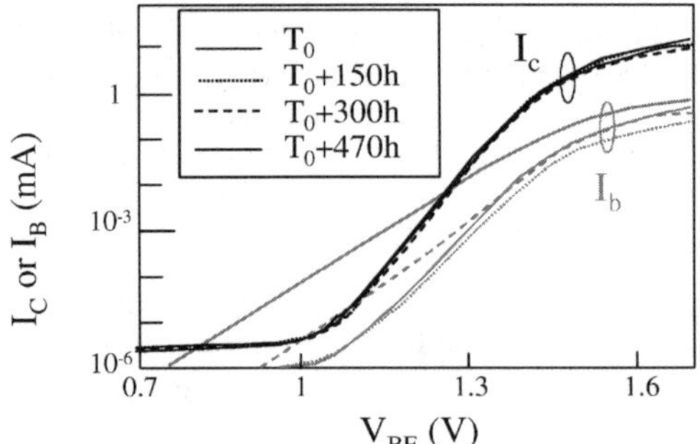

Figure 4: Gummel-plots evolution (2B device) versus the stress time.

From the evolution of the LF noise spectral density S_V with frequency (figure 5), we can notice a Lorentzian noise component (at 10 kHz) superimposed to the flicker 1/f noise source. This Lorentzian contribution diminishes with the stress time in the early hours, as β increases with the surface state improvement between the GaInP passivation ledge and the extrinsic BE layer. However, the 1/f component remains constant during the stress.

Current noise spectral density S_I only features a 1/f noise source, and the evolution of its amplitude is reported on figure 6. The burn-in effect is noticeable on the improvement of the S_I noise source in the first hours of stress. This burn-in is strongly dependant on the quality of the interface (surface cleaning material and procedure used as well as the composition of the passivation layer [1]). The increase of S_I is once again relevant with the $I_{R\text{-}SCR}$ increase. From dynamic S parameters measurements, we also found a decrease on the current gain of the devices.

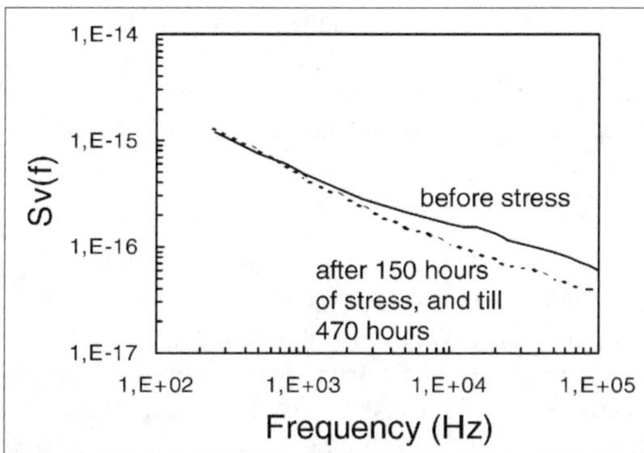

Figure 5: S_V LF noise spectrum versus frequency, before and after the 470 hours of stress (2B device, biased at V_{CE}=2V, I_C=3mA)..

Numerous studies have taken interest in the hydrogen reactions (under various forms, ionic, molecular,…) in semiconductors layers [6]: most of the H atoms are assimilated during the growing of the GaAs layers (hydrogen is used as a gas carrier for arsine under the AsH_3 form).

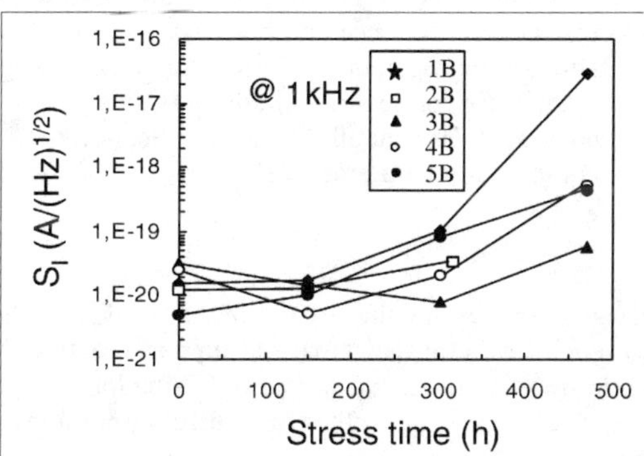

Figure 6: S_I LF noise spectral density at 1 kHz versus the stress time (#B lot devices, biased at V_{CE}=2V, I_C=3mA).

Measurements by DLTS on GaInP materials have been performed by the LPSC laboratory (Meudon, France), concluding in the presence of high concentration of hydrogen in this layer. The problem is thus attributed to the presence (high concentration up to few $10^{19}cm^{-3}$) of hydrogen combining with donors (C) in the semiconductor layers during the technological process: H diffuses in the base layer where it neutralizes the carbon donors forming C-H complexes. Under the combined action of the temperature and of the carrier's density, C-H complexes break and the base doping level is increased. As a consequence of the increase in the base doping, the current gain β decreases (figure 3). Moreover, the liberated hydrogen atoms generate a flux that affect the interface and surface states of the devices, and the current noise sources strongly increase (I_{R-SCR} and $I_{bI,e}$).

C-H breaking affects the surface states at the BE junction and the piezo-electric tensors: this results in a modification of the base leakage currents at the BE peripheral zone as seen on figure 3 ($I_{bl,e}$). Equations linking the threshold voltage (and also the current gain) to the doping level of the base have been used to correlate the measured V_T variations (also β variations) to the N_D increase: 60% variation on N_D has been found.

CONCLUSIONS

Stress conditions have been applied to HBTs devices: static and low-frequency noise measurements (and dynamic S parameters measurements) have been realized in order to give a physical interpretation about the electrical parameters (and electrical performances) degradation. These failures originate from hydrogen reactions with the GaAs layers. The exact mechanism by which hydrogen degrades the device's performances and by which hydrogen reaches the active area are still not exactly known and the investigations are still under progress. However, static and low frequency noise measurements have evidenced the part that hydrogen takes in the fluctuations of β with time under moderate junction temperature stress. The phenomenon involves complex C-H breaking, affecting the effective doping level of the base layer while H diffusion leads to modifications in the leakage currents on the base. The study on HBT devices is not trite (in comparison with TLM structures, or semiconductor layers) because the measured electrical performances are related to numerous parameters such as the transit time (emitter, base or collector), the injection efficiency, access resistances, interface or surface traps, … However, this kind of study is necessary to get the signature of hydrogen on the failure mode occurring on these complexes structures. This experimental approach for devices reliability is thus complementary with the specific studies on material layers that allow the identification of hydrogen and its chemical form (H_2, H_+, H_-, combination with other elements).

ACKNOWLEDGEMENT

The authors would like to thank S. Delage for providing the devices of this work, and the CNES (French spatial agency) and DGA (French DoD) for their financial support (contract N° 844/96/CNES/0512.00). We also want to acknowledge the LPSC laboratory (Meudon, France) for their participation to this program, and specially Pr. J. Chevallier for fruitful discussions.

REFERENCES

[1] M. Borgarino, J.G. Tartarin, S. Delage, R. Plana, F. Fantini, J. Graffeuil, 'Correlation between the burn-in effect and the extrinsic base surface quality in C-doped GaInP/GaAs HBTs', GAAS98 symposium, Amsterdam, pp. 301-306.
[2] T. Henderson, 'Modeling gallium arsenide heterojunction bipolar transistor ledge variations for insight into device reliability', Pergamon, micro. reliability 42, 2002, pp. 1011-1020
[3] S. Delage et al., 'The correlation between material properties and HBT reliability', GAAS99 symposium, Munich 99, pp. 246-251.
[4] T. Takahashi, S. Sasa, A. Kawano, T. Iwai, T. Fujii, 'High-reliability InGaP/GaAs HBTs fabricated by self-aligned process', IEDM 94, pp191-194.
[5] J.J. Liou et al., 'Base and collector leakage currents of AlGaAs/GaAs heterojunction bipolar transistor', J. Appl . Phys. 76 (5), 1 September 1994.
[6] D;M Kozuch, M.Stavola,'Passivation of carbon-doped GaAS layers by hydrogen introduced by annealing and growth ambients', J. Applied physics, 73 (8), April 93, pp. 3716-3724.

AUTHOR INDEX

DATE DUE

Printed
in USA

HIGHSMITH #45230